Springer Theses

Recognizing Outstanding Ph.D. Research

For further volumes:
http://www.springer.com/series/8790

Aims and Scope

The series "Springer Theses" brings together a selection of the very best Ph.D. theses from around the world and across the physical sciences. Nominated and endorsed by two recognized specialists, each published volume has been selected for its scientific excellence and the high impact of its contents for the pertinent field of research. For greater accessibility to non-specialists, the published versions include an extended introduction, as well as a foreword by the student's supervisor explaining the special relevance of the work for the field. As a whole, the series will provide a valuable resource both for newcomers to the research fields described, and for other scientists seeking detailed background information on special questions. Finally, it provides an accredited documentation of the valuable contributions made by today's younger generation of scientists.

Theses are accepted into the series by invited nomination only and must fulfill all of the following criteria

- They must be written in good English.
- The topic should fall within the confines of Chemistry, Physics, Earth Sciences, Engineering and related interdisciplinary fields such as Materials, Nanoscience, Chemical Engineering, Complex Systems and Biophysics.
- The work reported in the thesis must represent a significant scientific advance.
- If the thesis includes previously published material, permission to reproduce this must be gained from the respective copyright holder.
- They must have been examined and passed during the 12 months prior to nomination.
- Each thesis should include a foreword by the supervisor outlining the significance of its content.
- The theses should have a clearly defined structure including an introduction accessible to scientists not expert in that particular field.

Mathew Richard Bullimore

Scattering Amplitudes and Wilson Loops in Twistor Space

Doctoral Thesis accepted
by the University of Oxford, UK

Author
Dr. Mathew Richard Bullimore
Mathematical Institute
Radcliffe Observatory Quarter
University of Oxford
Oxford
UK

Foreword by
Prof. Lionel J. Mason
Mathematical Institute
University of Oxford
Oxford
UK

ISSN 2190-5053 ISSN 2190-5061 (electronic)
ISBN 978-3-319-34618-2 ISBN 978-3-319-00909-4 (eBook)
DOI 10.1007/978-3-319-00909-4
Springer Cham Heidelberg New York Dordrecht London

Printed on acid-free paper

Springer is part of Springer Science+Business Media (www.springer.com)

Supervisor's Foreword

It is a pleasure to write this foreword for Mat Bullimore's thesis. This research took place at an exciting time for the development of the subject and he was able to not only benefit from a collaboration with people in Oxford, but perhaps more importantly with David Skinner who had recently left Oxford for the Perimeter Institute and who played the role of an additional supervisor. There was also much fruitful interaction with the group led by Nima Arkani-Hamed and Freddy Cachazo in the IAS Princeton and Perimeter respectively. Of course this rapid development meant that, although there were opportunites, there was much for a new student to learn and it was necessary to be very quick to keep up with the competition.

Scattering amplitudes have always been a key output of quantum field theories. They are required in order to understand the physical consequences of particle models and to enable their predictions to be compared to experiment. Their study has become particularly pressing with the wealth of data arriving from the Large Hadron Collider for testing the standard model. However, the last 10 years has seen a completely different emphasis in the study of scattering amplitudes, arising from the unexpected mathematical structures that have been discovered that cry out for new physical and mathematical insights into the origins of physical theories.

Hints of such structures were first observed some time ago when some of the most elementary amplitudes, the so-called MHV amplitudes, turned out to have remarkably simple formulae despite the complexity of their origin as a sum of Feynman diagrams. More recently Witten showed that tree-level Yang-Mills amplitudes have a remarkably simple structure in twistor space and indeed could be understood as arising from a string theory in twistor space (as opposed to the more usual one in space-time). Twistor space is an auxilliary space that encapsulates the degrees of freedom of spinning massless particles in a complex three-dimensional projective space. This original twistor string was not viable as a model for conventional physics owing to the presence of conformal gravity. Nevertheless, the observation remained that, in twistor space, formulae for amplitudes are somehow simpler and more string like, being supported on curves of various degrees in twistor space.

There were a number of important outputs of these ideas. They included the MHV formalism, a particularly efficient Feynman diagram-like formalism for gauge theory amplitudes. Indeed, it had already emerged that the MHV formalism is indeed the Feynman diagram formalism for a gauge theory action principle in twistor space that has been expressed in an axial gauge. Another important set of tools are BCFW recursion relations (due to Britto, Cachazo, Feng and Witten) that determine more complicated amplitudes in terms of easier ones.

A major conjecture that had come to the fore in the run up to Mat's thesis was the surprising amplitude/Wilson duality: that planar amplitudes for N = 4 super Yang-Mills could computed as the correlation function of a Wilson loop. This had emerged from a string T-duality applied in the context of AdS/CFT that had initially seemed far fetched, but which worked out remarkably well in examples when details were worked out. Planar N = 4 super Yang-Mills became the optimal playground for trying to understand the mathematical structure of amplitudes.

The big open questions in these twistor related approaches was as to how one might study loop amplitudes and more general correlation functions using twistor theory. A halfway house between tree and loop ampltiudes are the so-called leading singularlities of loop amplitudes. A remarkable twistor-related generating formula for these had been obtained by the Princeton-Perimeter Institute collaboration in terms of a contour integral in a Grassmannian. His first paper gave a detailed analysis of the string-like structure of leading singularities in twistor space and the closely related expressions arising from BCFW recursion in twistor space, and the Grassmannian formulae was shown to arise partially as a consequence of such a twistor-string representation. This led to a classification scheme for leading singularities.

The next year saw further major developments in which it was seen how to gain a firmer grip on loop amplitudes via its integrand. On the one hand the Princeton/Perimeter collaboration were able to generate the all-loop integrand via an extension of BCFW recursion, whereas on the other, Mat with collaborators in Oxford and Perimeter was able to generate it using the MHV formalism in twistor space. This led to the realization that the amplitude/Wilson-loop duality could be expressed and proved perturbatively at the level of the integrand in twistor space from the twistor action. The Wilson loop was seen to correspond to holomorphic link invariants in twistor space and Mat, with David Skinner was able to show how the amplitude/Wilson-loop correspondence could be seen to follow from the loop equations adapted to holomorphic linking, a complex analytic generalization of real link invariants in three space.

The next year saw further important developments. One, not covered in the thesis but that Mat was involved with, was the observation that the twistor action framework was able to compute essentially arbitrary observables (although some with greater ease than others), and this allowed the proof of a correspondence between certain limits of certain correlation functions and null polygonal Wilson-loops, and hence planar amplitudes. The other was the study of symmetries and anomalies of the Wilson-loop. This led to certain descent equations that in principle allow the construction of integrated amplitudes. This was a great breakthrough as previously the methods had been restricted to computing integrands for correlation functions and amplitudes at loop level, but the anomalies led to descent equations that in principle at least, determines the integrated amplitude directly, and that could be used to compute highly nontrivial examples.

The subject is still evolving rapidly and promises many further new insights, but this thesis will remain a valuable resource, providing a detailed introduction and recording many of the key insights of the last few years.

Oxford, September 2013 Prof. Lionel J. Mason

the field has previously [illegible] has been restricted to [illegible] [illegible] [illegible] level, but [illegible] led to [illegible] that [illegible] the integrated [illegible] directly, and that could [illegible] important example.

The [illegible] will evolve rapidly and [illegible] further new insights [illegible] and [illegible] of the [illegible] in years of [illegible].

[illegible] [illegible]

Abstract

Scattering amplitudes are fundamental and remarkably rich observables in quantum field theory. The basic observation that makes scattering amplitudes fascinating is that in theories with massless particles of spin $s \geq 1$, they are much simpler than might be expected from traditional Feynman diagram techniques for computing them. This simple observation might ultimately have profound consequences for our view of quantum field theory. The broad aim of this thesis is to understand and exploit the hidden simplicity and structure in scattering amplitudes.

The quantum field theory with the simplest scattering amplitudes in four dimensions is planar $\mathcal{N} = 4$ supersymmetric Yang–Mills theory. This theory has provided considerable inspiration in developing new computational techniques and has provided many important theoretical insights. In this theory, there is a remarkable correspondence between scattering amplitudes and null polygonal Wilson loops, observables which on first inspection look very different. In this thesis, we will provide new insights into this correspondence using methods from twistor theory and exploit the symmetries of the problem to find new ways of computing scattering amplitudes.

Acknowledgements

This thesis is dedicated to my parents, Ann and Kevin Bullimore, for nurturing my interest in science and mathematics. I would like to thank them for their love and support throughout my studies. I would also like to express my heartfelt appreciation to Kellie Jacques for her love and patience.

I would like to thank my supervisor, John March-Russell, for encouraging me to undertake research in fundamental physics and for motivating me to think about the wider implications of my work. I am especially grateful to Lionel Mason and David Skinner who guided me into the world of research – this thesis would not have been possible without their ideas, insight and advice. I would also particularly like to thank Andrew Hodges, whose ideas have provided significant motivation and inspiration for the work in this thesis. Finally, I am grateful to the great many friends and physicists with whom I have interacted during my studies - I have learnt a tremendous amount from them.

Contents

Chapter 1
Introduction

Scattering amplitudes are fundamental and remarkably rich observables in quantum field theory. The single most important observation in this subject is that scattering amplitudes are often much simpler than one expects from a typical Feynman diagram expansion. Understanding the implications of this simple fact has lead to the development of new and efficient computational techniques and to the discovery of new mathematical structures in theoretical physics.

The observation is most striking for scattering amplitudes in theories of massless particles with spin $s \geq 1$. The reason is that in order to construct a local quantum field theory description for such particles, one is forced to introduce gauge redundancy into the lagrangian formulation, leading to significantly more complicated Feynman rules. On the other hand, scattering amplitudes are not local observables. Indeed, scattering amplitudes in such theories possess surprising simplicity and structure that is not apparent in the lagrangian.

The above considerations lead one to attempt to compute scattering amplitudes of a given helicity directly using fundamental principles based on unitarity and factorization. An important tool in this endeavor is the introduction of the spinor helicity formalism, where massless momenta are decomposed into spinors, $p^{\alpha\dot{\alpha}} = \lambda^{\alpha}\tilde{\lambda}^{\dot{\alpha}}$, and scattering amplitudes transform with weight

$$A_{h_1 \ldots h_n} \to t^{-\sum_i 2h_i} A_{h_1 \ldots h_n} \tag{1.1}$$

under little group transformations $(\lambda_i, \tilde{\lambda}_i) \to (t\lambda_i, t^{-1}\tilde{\lambda}_i)$. This decomposition allowed Parke and Taylor to find a stunningly simple formula for the tree-level scattering of six incoming gluons in a particular helicity combination [1],

$$A_{--++++} = \frac{\langle 12 \rangle^4}{\langle 12 \rangle \langle 23 \rangle \ldots \langle 61 \rangle} \tag{1.2}$$

with an immediate generalization to any number of particles. This simple formula is the summation of two hundred and twenty Feynman diagrams. This simplification

M. R. Bullimore, *Scattering Amplitudes and Wilson Loops in Twistor Space*, Springer Theses, DOI: 10.1007/978-3-319-00909-4_1,

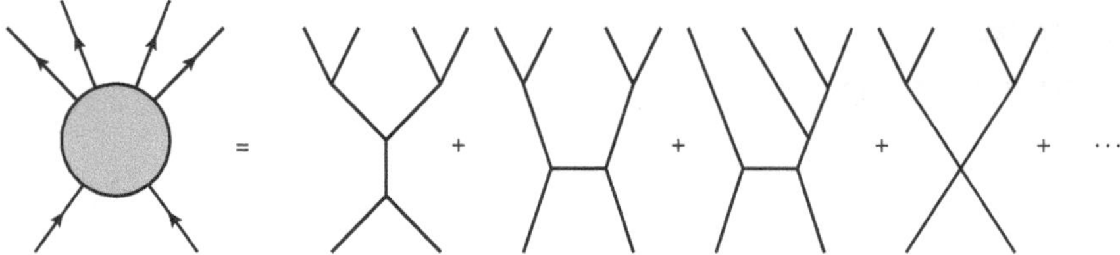

Fig. 1.1 The scattering of six gluons requires summing two-hundred and twenty Feynman diagrams representing intermediate interactions that can take place

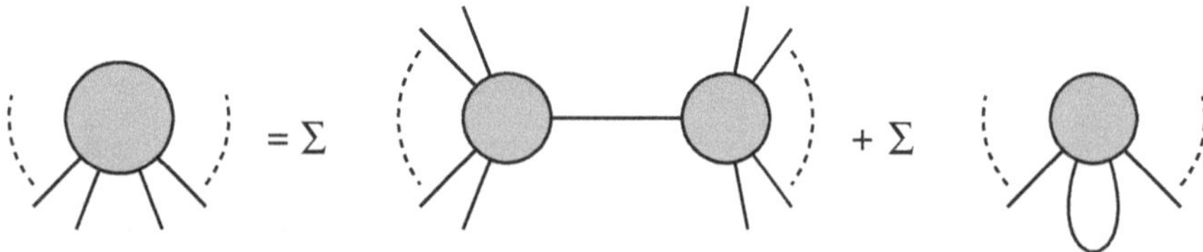

Fig. 1.2 A graphical representation of on-shell recursion relations for computing planar scattering amplitudes

is just a glimpse of the remarkable structure in scattering amplitudes in gauge and gravitational theories (Fig. 1.1).

Much progress has been made in computational efficiency and structural understanding of scattering amplitudes with the introduction of on-shell recursion relations for tree-level amplitudes [2, 3], MHV diagram expansions [4, 5] and unitarity based methods for loop amplitudes [6–8]. Such methods are characterized by constructing amplitudes in a way that remains completely on-shell, and are thus greatly more efficient than the traditional Feynman diagram expansion (Fig. 1.2).

When looking for new theoretical structures and computational techniques, one first examines the simplest theories and then extends the lessons learned to more complicated examples. The quantum field theory with the simplest scattering amplitudes in four dimensions is the maximally supersymmetric $\mathcal{N} = 4$ Yang-Mills theory. This theory has provided significant inspiration in developing new computational techniques and has provided many important theoretical insights. This quantum field theory is also of purely theoretical interest because of the remarkable duality with type IIB string theory in $AdS_5 \times S^5$ [9–11] and has lead to surprising connections with other areas of mathematical physics.

An important consequence of AdS / CFT duality is the correspondence between of planar scattering amplitudes and null polygonal Wilson loops. This remarkable correspondence was first discovered at strong coupling [12] and subsequently observed in perturbative calculations at weak coupling [13–16]. The amplitude/ Wilson loop correspondence inspired the discovery of a hidden superconformal symmetry of scattering amplitudes in planar $\mathcal{N} = 4$ supersymmetric Yang-Mills theory, named *dual* superconformal symmetry [17]. This additional symmetry can be understood as a consequence of the invariance of IIB string theory on $AdS_5 \times S^5$ under

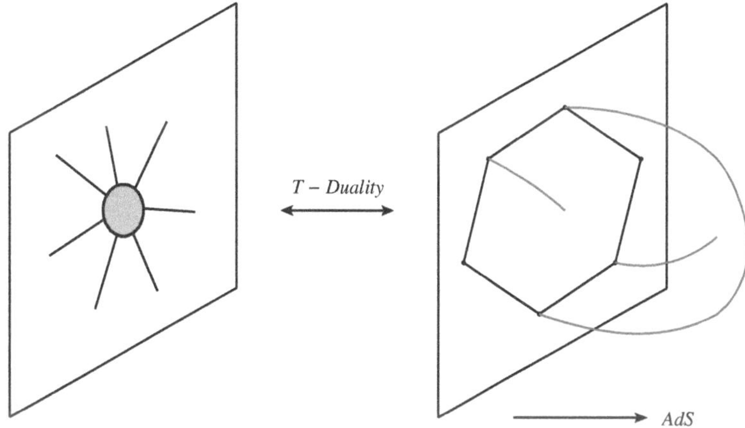

Fig. 1.3 A schematic illustration of the amplitude—Wilson loop correspondence realized at strong coupling as T-duality in AdS/CFT duality

T-duality [18, 19]. The hidden symmetries (or more precisely their anomalous behavior) are extremely powerful and are conjectured to determine the complete perturbative S-matrix of the theory (Fig. 1.3).

In this thesis, we apply the methods of twistor theory to understand the simplicity and structure of gauge theory scattering amplitudes. Twistor theory is a framework where the conformal structure of space-time is reformulated in terms of complex geometry, allowing remarkable geometric insight into physical problems involving massless particles [20, 21]. In particular, twistor theory has provided important insights into the theory of scattering amplitudes since its inception, see for example [21], and the twistor diagram program [22–24]. The application of twistor space methods to scattering amplitudes was given new impetus in the seminal paper on twistor string theory [25] and subsequent developments [26–28]. An important tool in this thesis is the formulation of an action for supersymmetric Yang-Mills theories in twistor space [29, 30], allowing certain perturbative quantum field theory calculations to be perfomed directly in twistor space, in place of more complicated calculations in space-time (Fig. 1.3).

In this thesis, we almost exclusively work in 'momentum twistor space', which is the twistor space associated to the space-time of the null polygonal Wilson loop [31]. Momentum twistors completely solve all kinematic constraints for planar amplitudes of massless particles. In addition, they manifest dual superconformal symmetry and have greatly enhanced our understanding of the scattering amplitude / Wilson loop correspondence, allowing the discovery of a supersymmetric extension of the Wilson loop that describes scattering amplitudes beyond the simplest MHV sector [32, 33].

Let us now summarize the content of the remaining chapters: Chap. 2 contains an introduction to twistor space geometry and quantum field theory in twistor space, and in particular the twistor space formulation of $\mathcal{N} = 4$ supersymmetric Yang-Mills theory. This chapter is intended as a review and simply presents the relevant material in a way that is maximally orientated towards applications to scattering amplitudes and Wilson loops. On the other hand, subsequent chapters present original work by the author, including both individual work and work performed in collaboration with colleagues.

Chapter 3 presents a supersymmetric version of the MHV diagram expansion for planar $\mathcal{N} = 4$ supersymmetric Yang-Mills theory and reformulates it in momentum twistor space. This expansion provides a particularly simple and efficient method to construct tree-level superamplitudes and loop integrands of the theory in a way that manifests dual superconformal symmetry. This chapter is based on reference [34] performed in collaboration with David Skinner and Lionel Mason.

Chapter 4 formulates the 'all-line' recursion relation in momentum twistor space. The recursion relation is completely solved and shown to generate the supersymmetric MHV diagram expansion presented in Chap. 3, thus proving its validity for all planar tree-level superamplitudes and loop integrands. This chapter is based on reference [35] of which I am the sole author.

Chapter 5 first reviews the supersymmetric extension of the amplitude/Wilson loop correspondence, as formulated in momentum twistor space, and its relationship with the MHV diagram expansion [32]. This is included for completeness and is not original work by the author. In the second part, we derive a holomorphic analogue of the loop equations, describing how the momentum twistor Wilson loop behaves under holomorphic deformations, and use it to independently derive on-shell recursion relations. This proves the amplitude/Wilson loop correspondence for tree-level superamplitudes and four-dimensional loop integrands. The second part is based on the author's work [36] in collaboration with David Skinner.

Chapter 6 considers the breaking of dual superconformal symmetry. We explain why one chiral half of the dual supersymmetry generators are broken in loop amplitudes, even for observables with no infrared divergences, and derive the anomalous Ward identities. The anomalous Ward identity is then re-expressed at the level of the kinematic data by integrating out an additional particle. The resulting equation is recursive and places a powerful constraint on the planar S-matrix. This chapter is based on the author's work [37] in collaboration with David Skinner.

References

1. S.J. Parke, T. Taylor, An amplitude for *n* gluon scattering. Phys. Rev. Lett. **56**, 2459 (1986)
2. R. Britto, F. Cachazo, B. Feng, New recursion relations for tree amplitudes of gluons. Nucl. Phys. **B715**, 499–522 (2005). [hep-th/0412308]
3. R. Britto, F. Cachazo, B. Feng, E. Witten, Direct proof of tree-level recursion relation in Yang-Mills theory. Phys. Rev. Lett. **94**, 181602 (2005). [hep-th/0501052]
4. F. Cachazo, P. Svrcek, E. Witten, MHV vertices and tree amplitudes in gauge theory. J. High Energy Phys. **0409**, 006 (2004). [hep-th/0403047]
5. K. Risager, A direct proof of the CSW rules. J. High Energy Phys. **0512**, 003 (2005). [hep-th/0508206]
6. Z. Bern, L.J. Dixon, D.C. Dunbar, D.A. Kosower, One loop n point gauge theory amplitudes, unitarity and collinear limits. Nucl. Phys. **B425**, 217–260 (1994). [hep-ph/9403226]

7. Z. Bern, L.J. Dixon, D.C. Dunbar, D.A. Kosower, Fusing gauge theory tree amplitudes into loop amplitudes. Nucl. Phys. **B435**, 59–101 (1995). [hep-ph/9409265]
8. C. Berger, Z. Bern, L. Dixon, F. Febres Cordero, D. Forde et al., An automated implementation of on-shell methods for one-loop amplitudes. Phys. Rev. **D78**, 036003 (2008). [arXiv:0803.4180]
9. J.M. Maldacena, The large N limit of superconformal field theories and supergravity. Adv. Theor. Math. Phys. **2**, 231–252 (1998). [hep-th/9711200]
10. E. Witten, Anti-de Sitter space and holography. Adv. Theor. Math. Phys. **2**, 253–291 (1998). [hep-th/9802150]
11. S. Gubser, I.R. Klebanov, A.M. Polyakov, Gauge theory correlators from noncritical string theory. Phys. Lett. **B428**, 105–114 (1998). [hep-th/9802109]
12. L.F. Alday, J.M. Maldacena, Gluon scattering amplitudes at strong coupling. J. High Energy Phys. **0706**, 064 (2007). [arXiv:0705.0303]
13. G. Korchemsky, J. Drummond, E. Sokatchev, Conformal properties of four-gluon planar amplitudes and Wilson loops. Nucl. Phys. **B795**, 385–408 (2008). [arXiv:0707.0243]
14. J. Drummond, J. Henn, G. Korchemsky, E. Sokatchev, Conformal ward identities for Wilson loops and a test of the duality with gluon amplitudes. Nucl. Phys. **B826**, 337–364 (2010). [arXiv:0712.1223]
15. J. Drummond, J. Henn, G. Korchemsky, E. Sokatchev, On planar gluon amplitudes/Wilson loops duality. Nucl. Phys. **B795**, 52–68 (2008). [arXiv:0709.2368]
16. A. Brandhuber, P. Heslop, G. Travaglini, MHV amplitudes in N=4 super Yang-Mills and Wilson loops. Nucl. Phys. **B794**, 231–243 (2008). [arXiv:0707.1153]
17. J. Drummond, J. Henn, G. Korchemsky, E. Sokatchev, Dual superconformal symmetry of scattering amplitudes in N=4 super-Yang-Mills theory. Nucl. Phys. **B828**, 317–374 (2010). [arXiv:0807.1095]
18. N. Berkovits, J. Maldacena, Fermionic T-Duality, dual superconformal symmetry, and the amplitude/Wilson loop connection. J. High Energy Phys. **0809**, 062 (2008). [arXiv:0807.3196]
19. N. Beisert, T-Duality, dual conformal symmetry and integrability for strings on adS(5) x S**5. Fortsch. Phys. **57**, 329–337 (2009). [arXiv:0903.0609]
20. R. Penrose, Twistor algebra. J. Math. Phys. **8**, 345 (1967)
21. R. Penrose, M.A. MacCallum, Twistor theory: An approach to the quantization of fields and space-time. Phys. Rept. **6**, 241–316 (1972)
22. A.P. Hodges, S. Huggett, Twistor diagrams. Surveys High Energ. Phys. **1**, 333–353 (1980)
23. A.P. Hodges, Twistor diagrams and massless Moller scattering. Proc. Roy. Soc. Lond. **A385**, 207–228 (1983)
24. A.P. Hodges, Twistor diagrams and massless compton scattering. Proc. Roy. Soc. Lond. **A386**, 185–210 (1983)
25. E. Witten, Perturbative gauge theory as a string theory in twistor space. Commun. Math. Phys. **252**, 189–258 (2004). [hep-th/0312171]
26. N. Berkovits, An alternative string theory in twistor space for N=4 superYang-Mills. Phys. Rev. Lett. **93**, 011601 (2004). [hep-th/0402045]
27. N. Berkovits, E. Witten, Conformal supergravity in twistor-string theory. J. High Energy Phys. **0408**, 009 (2004). [hep-th/0406051]
28. R. Roiban, M. Spradlin, A. Volovich, On the tree level S matrix of Yang-Mills theory. Phys. Rev. **D70**, 026009 (2004). [hep-th/0403190]
29. L. Mason, Twistor actions for non-self-dual fields: A derivation of twistor-string theory. J. High Energy Phys. **0510**, 009 (2005). [hep-th/0507269]
30. R. Boels, L. Mason, D. Skinner, Supersymmetric Gauge theories in twistor space. J. High Energy Phys. **0702**, 014 (2007). [hep-th/0604040]
31. A. Hodges, Eliminating spurious poles from gauge-theoretic amplitudes, arXiv:0905.1473
32. L. Mason, D. Skinner, The complete planar S-matrix of N=4 SYM as a Wilson loop in twistor space. J. High Energy Phys. **1012**, 018 (2010). [arXiv:1009.2225]
33. S. Caron-Huot, Notes on the scattering amplitude/Wilson loop duality. J. High Energy Phys. **1107**, 058 (2011). [arXiv:1010.1167]

34. M. Bullimore, L. Mason, D. Skinner, MHV diagrams in momentum twistor space. J. High Energy Phys. **1012**, 032 (2010). [arXiv:1009.1854]
35. M. Bullimore, MHV diagrams from an all-line recursion relation, arXiv:1010.5921
36. M. Bullimore, D. Skinner, Holomorphic linking, loop equations and scattering amplitudes in twistor space, arXiv:1101.1329
37. M. Bullimore, D. Skinner, Descent equations for superamplitudes, arXiv:1112.1056

Chapter 2
Review

This chapter is an introduction to some aspects of twistor geometry and the twistor space description of space-time conformal field theories. In particular, we will focus on the twistor space description of $\mathcal{N} = 4$ maximally supersymmetric Yang-Mills theory. The material is not original. I hope only to provide a presentation of the material that is oriented towards recent applications to scattering amplitudes and Wilson loops. For a recent review with similar aspirations see reference [1]. Some additional useful resources on twistor theory are [2, 3]

2.1 Twistor Theory

2.1.1 Complexified Compactified Minkowski Space

Conformal field theories in four dimensions are naturally defined on the conformal compactification of Minkowski space, whose complexification will be denoted by $\mathbb{CM}^\sharp$, and on which the complexified conformal group, $\mathrm{SL}(4, \mathbb{C})$, naturally acts. The analytic continuation to complex space-time coordinates is important throughout this thesis. The reason is that the space-time coordinates will actually describe the four-momenta of scattering states, and the techniques that we use depend on the analytic properties of scattering amplitudes as functions of complex momenta. When reality conditions are important we will state this explicitly.

Complexified compactified Minkowski space, $\mathbb{CM}^\sharp$, has a beautiful description as a complex quadric surface in the complex projective space $\mathbb{CP}^5$. Introducing six homogeneous coordinates on $\mathbb{CP}^5$ transforming in the anti-symmetric tensor representation of $\mathrm{SL}(4, \mathbb{C})$,

$$X^{IJ} = -X^{JI} \qquad I, J = 1, \ldots, 4, \tag{2.1}$$

M. R. Bullimore, *Scattering Amplitudes and Wilson Loops in Twistor Space*, Springer Theses, DOI: 10.1007/978-3-319-00909-4_2,

the quadric surface is

$$X \cdot X := X_{IJ}\, X^{IJ} = 0. \tag{2.2}$$

The anti-symmetric pair of indices has been lowered using the canonical anti-symmetric tensor ϵ_{IJKL} of the complexified conformal group SL(4, $\mathbb{C}$),

$$X_{IJ} = \frac{1}{2}\epsilon_{IJKL}X^{KL}. \tag{2.3}$$

The conformal structure of complexified compactified Minkowski space $\mathbb{CM}^\sharp$ is then inherited from the anti-symmetric tensor ϵ_{IJKL}: two points X and Y are null separated if and only if

$$X \cdot Y = 0. \tag{2.4}$$

Allowing all possible Y, the above equation defines the tangent plane to the quadric at X, and hence its intersection with the quadric is the complex null cone of X.

2.1.2 Twistor Space

Twistor space is the complex projective space $\mathbb{CP}^3$, whose four complex homogeneous coordinates $Z^I \sim rZ^I$ transform in the fundamental representation of the complexified conformal group SL(4, $\mathbb{C}$). Strictly speaking, the complexified conformal group is the group of complex projective linear transformations PGL(4, $\mathbb{C}$) = SL(4, $\mathbb{C}$)/$\mathbb{Z}_4$ acting on $\mathbb{CP}^3$.

The relationship between twistor space geometry and space-time conformal structure is encapsulated in the so-called 'incidence relation'. Given a point in complexified compactified space-time X, the incidence relation is

$$X_{IJ}\, Z^J = 0. \tag{2.5}$$

The condition $X \cdot X = 0$ implies that the skew tensor X^{IJ} has rank two, so there is only one independent equation up to complex rescalings. Thus the incidence relation defines a holomorphically embedded line in twistor space, $X = \mathbb{CP}^1$. The skew tensor X^{IJ} can be reconstructed from the line by forming

$$X^{IJ} = A^{[I}B^{J]} \tag{2.6}$$

where A^I and B^I are any two points on the line X. Conversely, any complex projective line in twistor space can be expressed as above, thus defining a point in $\mathbb{CM}^\sharp$. Hence points in space-time are in one-to-one correspondence with complex projective lines in twistor space Fig. 2.1.

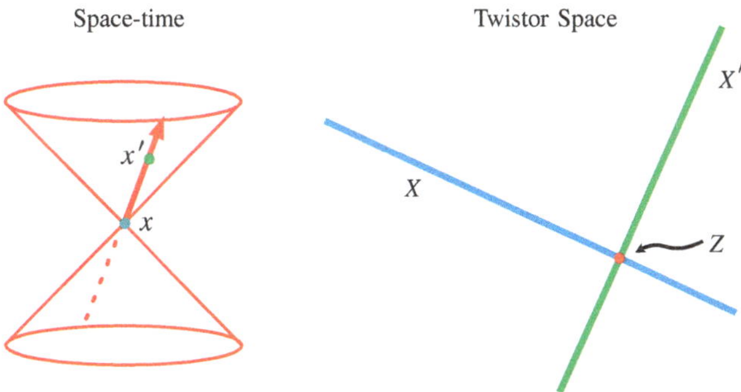

Fig. 2.1 Points in space-time correspond to lines in twistor space. Two space-time points are null separated if and only if their corresponding twistor lines intersect

A fundamental property of this correspondence is that two space-time points are null separated if and only if the corresponding twistor lines intersect intersect. To see this, suppose that there is a simultaneous solution of the equations

$$X_{IJ}\,Z^J = 0 \qquad Y_{IJ}\,Z^J = 0 \tag{2.7}$$

then

$$X^{IJ} = Z^{[I}A^{J]} \qquad Y^{IJ} = Z^{[I}B^{J]} \tag{2.8}$$

for some A^I and B^J and consequently $X \cdot Y = 0$. The converse statement is similarly straightforward. The complex structure of twistor space is therefore equivalent to the conformal structure of compactified space-time.

For a given twistor Z^I, the general solution of the incidence relation is

$$X^{IJ} = X_0^{IJ} + Y^{[I}Z^{J]}, \tag{2.9}$$

where X_0 is any particular solution and Y^I is an arbitrary twistor. The coordinates Y^I thus parametrize a complex two-dimensional plane in $\mathbb{CM}^\sharp$ all of whose tangent vectors are mutually orthogonal and null. This is called an 'α-plane'. Thus points in twistor space correspond to α-planes in space-time.

2.1.3 The Infinity Twistor

For many purposes, it is convenient to break manifest conformal symmetry and introduce a local affine coordinate system on Minkowski space. This requires choosing a point at infinity, represented by a simple anti-symmetric tensor I^{IJ}. The points in $\mathbb{CM}^\sharp$ lying in the surface

$$\mathcal{I} = \{X \in \mathbb{CM}^\sharp,\ I \cdot X = 0\} \tag{2.10}$$

define the light-cone of the point at infinity. Removing these points from compactified complexified space-time, we obtain complexified Minkowski space $\mathbb{CM} = \mathbb{CM}^\sharp \backslash \mathcal{I}$, which is isomorphic to $\mathbb{C}^4$. The point at infinity I^{IJ} defines a metric on the remaining $\mathbb{CM}$ defined by the formula

$$g(X, Y) = \frac{X \cdot Y}{(I \cdot X)\,(I \cdot Y)} \tag{2.11}$$

and breaking the complexified conformal symmetry SL(4, $\mathbb{C}$) down to the complexified Poincare group.

The fundamental indices of SL(4, $\mathbb{C}$) can now be decomposed into components transforming in spinor representations of the complexified Lorentz subgroup SL(2, $\mathbb{C}$)× SL(2, $\mathbb{C}$) in the following way,

$$Z^I = (\lambda_\alpha, \mu^{\dot\alpha}). \tag{2.12}$$

The point at infinity can be represented by the following simple matrices

$$I^{IJ} = \begin{pmatrix} 0 & 0 \\ 0 & \epsilon^{\dot\alpha\dot\beta} \end{pmatrix} \qquad I_{IJ} = \begin{pmatrix} \epsilon^{\alpha\beta} & 0 \\ 0 & 0 \end{pmatrix} \tag{2.13}$$

and then we can introduce affine coordinates $\{x^{\alpha\dot\alpha}\}$ on $\mathbb{CM}$ by

$$X^{IJ} = \begin{pmatrix} \epsilon_{\alpha\beta} & -ix_\alpha{}^{\dot\beta} \\ ix^{\dot\alpha}{}_\beta & -\frac{1}{2}x^2\epsilon^{\dot\alpha\dot\beta} \end{pmatrix}. \tag{2.14}$$

They are related to the standard affine coordinates $\{x^\mu\}$ on $\mathbb{C}^4$ through the linear relationship $x^{\alpha\dot\alpha} = x^\mu\sigma_\mu^{\alpha\dot\alpha}$. The metric (2.11) on complexified Minkowski space then becomes the standard metric on $\mathbb{C}^4$.

The twistors with vanishing primary components, $\lambda_\alpha = 0$, are incident on the light-cone at infinity $\mathcal{I}$, and removing these points from twistor space, we obtain a correspondence between lines in $\mathbb{CP}^3 - \{\lambda_\alpha = 0\}$ and points in $\mathbb{CM}$. The all important incidence relation describing the correspondence becomes

$$\mu^{\dot\alpha} = -ix^{\alpha\dot\alpha}\lambda_\alpha \tag{2.15}$$

and clearly defines a complex line in twistor space provided that $\lambda_\alpha \neq 0$. In the other direction, the general solution of the incidence relation for a given twistor is

$$x^{\alpha\dot\alpha} = x_0^{\alpha\dot\alpha} + \lambda^\alpha\sigma^{\dot\alpha} \tag{2.16}$$

where $x_0^{\alpha\dot{\alpha}}$ is one particular solution and $\sigma^{\dot{\alpha}}$ is an arbitrary spinor. This expression defines a totally null α-plane in affine coordinates.

Finally, we note that infinitessimal conformal transformations are now realized by the following generators

$$\begin{aligned} P_{\alpha\dot{\alpha}} &= \lambda_\alpha \frac{\partial}{\partial \mu^{\dot{\alpha}}} \\ J_{\alpha\beta} &= \frac{1}{2}\left(\lambda_\alpha \frac{\partial}{\partial \lambda^\beta} + \lambda_\beta \frac{\partial}{\partial \lambda^\alpha}\right) \qquad J_{\dot{\alpha}\dot{\beta}} = \frac{1}{2}\left(\mu_{\dot{\alpha}} \frac{\partial}{\partial \mu^{\dot{\beta}}} + \mu_{\dot{\beta}} \frac{\partial}{\partial \mu^{\dot{\alpha}}}\right) \\ D &= \frac{1}{2}\left(\lambda_\alpha \frac{\partial}{\partial \lambda_\alpha} - \mu^{\dot{\alpha}} \frac{\partial}{\partial \mu^{\dot{\alpha}}}\right) \qquad K^{\alpha\dot{\alpha}} = \mu^{\dot{\alpha}} \frac{\partial}{\partial \lambda_\alpha} \end{aligned} \tag{2.17}$$

in terms of the spinor components. However, we note that by following the integral curves of the above special conformal generator $K^{\alpha\dot{\alpha}}$, points at infinity can be reached at finite parameter values. In this sense, a complete realisation of finite conformal transformations requires the compactified space $\mathbb{CM}^\sharp$.

2.1.4 Reality Conditions

We now consider the reality condition appropriate for real Minkowski space, $\mathbb{M}$. The conformal group is now locally isomorphic to the real form SU(2, 2) preserving the pseudo-hermitian metric

$$g_{I\bar{J}} Z^I \bar{Z}^{\bar{J}} = \lambda_\alpha \bar{\mu}^\alpha + \mu^{\dot{\alpha}} \bar{\lambda}_{\dot{\alpha}}. \tag{2.18}$$

This partitions twistor space into three components

$$\mathbb{PT}^+ := \{Z \cdot \bar{Z} > 0\} \qquad \mathbb{PN} := \{Z \cdot \bar{Z} = 0\} \qquad \mathbb{PT}^- := \{Z \cdot \bar{Z} < 0\}. \tag{2.19}$$

For lines X lying entirely inside null twistor space $\mathbb{PN}$, the incidence relation and its complex conjugate imply that

$$0 = i(x - x^\dagger)^{\alpha\dot{\alpha}} \lambda_\alpha \bar{\lambda}_{\dot{\alpha}} \tag{2.20}$$

for all spinors λ^α. This condition is possible if and only if $x^{\alpha\dot{\alpha}}$ is hermitian and hence describes a real point in Minkowski space, $\mathbb{M}$. The line representing the point at infinity I is automatically inside $\mathbb{PN}$.

Consider now the α-plane associated to a point $Z^I \in \mathbb{PN}$. It is straightforward to show that the α-plane then necessarily contains a real point and conversely, that the existence of such a real point implies that $Z^I \in \mathbb{PN}$. The α-plane then intersects real Minkowski space in the null geodesic

$$x^{\alpha\dot{\alpha}} = x_0^{\alpha\dot{\alpha}} + r\lambda^\alpha\bar{\lambda}^{\dot{\alpha}} \tag{2.21}$$

where $x_0^{\alpha\dot{\alpha}} = \bar{\mu}^\alpha\mu^{\dot{\alpha}}/i\bar{\mu}^\alpha\lambda_\alpha$ and $r \in \mathbb{R}$, and furthermore, each null geodesic uniquely determines an α-plane. Thus, the points in null twistor space $\mathbb{PN}$ are in one-to-one correspondence with null geodesics in real Minkowski space $\mathbb{M}$.

2.1.5 The Penrose Transform

In this subsection, we describe the twistor space formulation of some free conformal field theories. Solutions of massless free field equations, for example,

$$\partial^{\alpha_1\dot{\alpha}_1}\phi_{\alpha_1\cdots\alpha_n} = 0 \qquad \text{or} \qquad \partial^{\alpha_1\dot{\alpha}_1}\phi_{\dot{\alpha}_1\cdots\dot{\alpha}_n} = 0 \tag{2.22}$$

are in one-to-one correspondence with cohomology classes on certain regions of twistor space. The concrete correspondence is provided by the Penrose transform [4, 5]. Here we proceed by example.

Let us begin by constructing solutions of the scalar wave equation

$$\Box\phi(x) = 0. \tag{2.23}$$

The starting point for the construction is an (0, 1)-form $\tilde{\phi}$ on some open region $U \subset \mathbb{CP}^3$, valued in sections of the line bundle $\mathcal{O}(-2)$. This means that $\tilde{\phi}$ has weight -2 in the homogeneous coordinates Z^I. It is important to consider an open region $U \subset \mathbb{CP}^3$, since there are no globally holomorphic sections of the line bundle $\mathcal{O}(-2)$. We can now pull $\tilde{\phi}$ back to the spinor bundle using the incidence relations and integrate over the fibres

$$\phi(x) = \int \langle\lambda\, \mathrm{d}\lambda\rangle \wedge \tilde{\phi}\,(\lambda_\alpha, ix^{\alpha\dot{\alpha}}\lambda_\alpha). \tag{2.24}$$

This expression defines a scalar field configuration on an open region of space-time corresponding to twistor lines inside $U \subset \mathbb{CP}^3$. The above integral expression is called the Penrose transform.

We now make two important observations. Firstly, the integral is invariant under the transformation

$$\tilde{\phi} \longrightarrow \tilde{\phi} + \bar{\partial}\,\tilde{\alpha} \tag{2.25}$$

where $\tilde{\alpha}$ is any section on $\mathcal{O}(-2)$ and hence depends only on the cohomology class in $H^{0,1}_{\bar{\partial}}(U, \mathcal{O}(-2))$ represented by $\tilde{\phi}$. Secondly, acting underneath the integral sign with the operator $\Box = \partial^{\alpha\dot{\alpha}}\partial_{\alpha\dot{\alpha}}$ the resulting expression vanishes since $\lambda^\alpha\lambda_\alpha = 0$.

Thus we have obtained a scalar field configuration obeying the scalar wave equation from a cohomology class $H^{0,1}_{\bar{\partial}}(U, \mathcal{O}(-2))$.

Let us consider a concrete example. A simple cohomology representative is the so called elementary state [6]

$$\tilde{\phi}(Z) = \frac{I^{IJ} P_I Q_J}{P_I Z^I} \bar{\partial} \frac{1}{Q_J Z^J}, \tag{2.26}$$

which is holomorphic away from the complex two-planes defined by $P := \{P_I Z^I = 0\}$ and $Q := \{Q_I Z^I = 0\}$. The Penrose transform is straightforwardly evaluated with the result

$$\phi(x) = \frac{1}{(x-y)^2}, \tag{2.27}$$

where the space-time point $y^{\alpha\dot{\alpha}} = p^\alpha q^{\dot{\alpha}} - q^\alpha p^{\dot{\alpha}}$ corresponds to the line formed by the intersection of the two-planes in twistor space, $Y = P \cap Q$. The Penrose transform is well-defined when the poles at the intersections $X \cap P$ and $X \cap Q$ do not coincide, or equivalently when the twistor lines X and Y are skew. If the lines X and Y intersect, then $(x-y)^2 = 0$ and the solution becomes singular.

This simple example provides some insight into the commonly used open regions for twistor space cohomology classes. Suppose that the intersection $P \cap Q$ lies entirely in $\mathbb{PT}^-$. Then the spacetime field configuration $\phi(x)$ is certainly non-singular on all lines $X \subset \mathbb{PT}^+$, or equivalently at spacetime points $x \in \mathbb{CM}^+$ that are future time-like. Such cohomology classes that are defined on the open region $U = \mathbb{PT}^+$ and lead to space-time field configurations that are non-singular on $\mathbb{CM}^+$ are called positive frequency fields. The analogous construction applies for negative frequency fields. It should be noted, however, that plane wave solutions are non-normalisable and cannot be accomodated in the present scheme.

In general, the Penrose transform provides an isomorphism between cohomology classes $H^{0,1}_{\bar{\partial}}(U, \mathcal{O}(2h-2))$ on some open region of twistor space $U \subset \mathbb{CP}^3$ and solutions of the zero-rest-mass field equations of helicity h. When the helicity $h \leq 0$ the integral transform becomes

$$\phi_{\alpha_1 \dots \alpha_h}(x) = \int_X \langle \lambda \mathrm{d}\lambda \rangle \lambda_{\alpha_1} \dots \lambda_{\alpha_{|h|}} \tilde{\phi}(\lambda_\alpha, i x^{\alpha\dot{\alpha}} \lambda_\alpha), \tag{2.28}$$

whereas when $h > 0$ we have

$$\phi_{\dot{\alpha}_1 \dots \dot{\alpha}_h}(x) = \int_X \langle \lambda \mathrm{d}\lambda \rangle \frac{\partial^h}{\partial \mu^{\dot{\alpha}_1} \dots \mu^{\alpha_h}} \tilde{\phi}(\lambda_\alpha, i x^{\alpha\dot{\alpha}} \lambda_\alpha). \tag{2.29}$$

In the following, we are mainly interested the description of non-abelian gauge fields, which require a more sophisticated framework in twistor space.

2.1.6 Self–Dual Yang–Mills Theory

We now consider the twistor space formulation of classical self-dual Yang-Mills theory. As above, we choose to work with Dolbeaut cohomology classes since this allows an off-shell formulation of the theory. This is essential when we come to constructing the twistor action and quantum mechanical perturbation theory in twistor space [7].

Our starting point is a smooth complex vector bundle of rank N

$$E \longrightarrow U \subset \mathbb{CP}^3 \tag{2.30}$$

with $c_1(E) = 0$ and equipped with an almost complex structure $\bar{D} = \bar{\partial} + a$. Later we will impose that the bundle be holomorphic on the open region $U \subset \mathbb{CP}^3$. However, it is important that we can construct a holomorphic frame on any line $X \subset U$ without imposing this condition.

The bundle is automatically holomorphic on pulling back to any line $X \subset U$ since $\bar{D}^2|_X = 0$ for dimensional reasons. It is also topologically trivial since our assumption $c_1(E) = 0$ implies also that $c_1(E|_X) = 0$. Hence, the bundle $E|_X$ is necessarily a sum of line bundles

$$E|_X = \bigoplus_i \mathcal{O}(d_i) \qquad \text{where} \qquad \sum_i d_i = 0 \tag{2.31}$$

and is holomorphically trivial if and only if $d_i = 0$ individually. In perturbation theory, we will always expand around a holomorphically trivial bundle, and small perturbations cannot change the discreet decomposition. Thus we can assume the bundle $E|_X$ to be holomorphically trivial.

This means we can find a smooth gauge transformation $h_X(\lambda, \bar{\lambda})$ such that

$$h_X^{-1} \circ (\bar{\partial} + a)\big|_X \circ h_X = \bar{\partial}\big|_X . \tag{2.32}$$

and under which covariantly holomorphic objects become simply holomorphic. Hence $h_X(\lambda, \bar{\lambda})$ is said to define a holomorphic frame for the bundle $E|_X$. The holomorphic frame itself obeys

$$(\bar{\partial} + a)\big|_X \, h_X = 0 \tag{2.33}$$

and hence is defined up to

$$h_X(\lambda, \bar{\lambda}) \to h_X(\lambda, \bar{\lambda})\, g(X) \tag{2.34}$$

where the gauge transformation $g(X)$ is globally holomorphic on the Riemann sphere and hence constant. We emphasize that the holomorphic frame can be constructed, at least in perturbation theory, without imposing that the bundle be holomorphic on the whole open region $U \subset \mathbb{CP}^3$. This observation will allow us to construct certain observables off-shell in twistor space.

In order to construct a space-time gauge bundle, no further conditions are required. Since the bundle $E|_X$ is automatically holomorphically trivial, we can always find N linearly independent globally holomorphic sections, which are unique up to constant GL(N) transformations. The space of such holomorphic sections $\Gamma(X, E|_X) \cong \mathbb{C}^{\mathrm{N}}$ thus provides the fibres of a complex vector bundle on an open region of space-time $\mathbb{CM}$ with complexified gauge group GL(N).

However, in order to construct a connection on the space-time gauge bundle, we must impose further conditions on the almost complex structure $\bar{D}$. We now pull the operator $\bar{D}$ back to the spin bundle using the projection map $p : \mathbb{S} \to U$ inherited from the incidence relations. The components are

$$\bar{D}_\lambda = \bar{\partial}_\lambda + a_\lambda \quad \bar{D}_{\dot{\alpha}} = \lambda^\alpha \partial_{\alpha\dot{\alpha}} + a_{\dot{\alpha}} \tag{2.35}$$

where the operators $\bar{\partial}_\lambda$ and $\lambda^\alpha \partial_{\alpha\dot{\alpha}}$ annihilate functions on the spin bundle $f(\lambda_\alpha, ix^{\alpha\dot{\alpha}}\lambda_\alpha)$ that are pulled back from holomorphic functions on U. They are extended to covariant operators using the components

$$a_\lambda = \bar{\partial}_\lambda \lrcorner\, p^* a \quad a_{\dot{\alpha}} = \lambda^\alpha \partial_{\alpha\dot{\alpha}} \lrcorner\, p^* a. \tag{2.36}$$

In this notation, the holomorphic frame obeys the equation $\bar{D}_\lambda\, h = 0$.

We have already mentioned that $\left[\bar{D}_\lambda, \bar{D}_\lambda\right] = 0$ for dimensional reasons. In particular, this allowed the construction of the holomorphic frame. In order to construct a space-time connection, we impose the further condition

$$\left[\bar{D}_\lambda, \bar{D}_{\dot{\alpha}}\right] = 0. \tag{2.37}$$

An immediate consequence, using the holomorphic frame condition $\bar{D}_\lambda h = 0$ twice, is that

$$\begin{aligned} \bar{\partial}_\lambda \left(h^{-1} \bar{D}_{\dot{\alpha}}\, h\right) &= h^{-1} \bar{D}_\lambda\, \bar{D}_{\dot{\alpha}}\, h \\ &= h^{-1}\left[\bar{D}_\lambda, \bar{D}_{\dot{\alpha}}\right] h \\ &= 0. \end{aligned} \tag{2.38}$$

Thus the combination $h^{-1} \bar{D}_{\dot{\alpha}}\, h$ is holomorphic along the fibres of the spin bundle and has weight one, and therefore must take the form

$$h^{-1} \bar{D}_{\dot{\alpha}}\, h = \lambda^\alpha\, A_{\alpha\dot{\alpha}}(x) \tag{2.39}$$

for some space-time field $A_{\alpha\dot{\alpha}}$. Under the gauge transformation (2.34) preserving the holomorphic frame we have $A_{\alpha\dot{\alpha}} \to g^{-1}(\partial_{\alpha\dot{\alpha}} + A_{\alpha\dot{\alpha}})g$, and hence we have obtained a gauge connection on the space-time complex vector bundle.

Finally, we impose the remaining condition

$$\left[\bar{D}_{\dot{\alpha}}, \bar{D}_{\dot{\beta}}\right] = 0 \tag{2.40}$$

meaning that the twistor bundle is now holomorphic on U. This equation implies that the curvature of the space-time connection $A_{\alpha\dot{\alpha}}$ is flat on restriction to any α-plane. This condition can equivalently be expressed

$$\begin{aligned}
\lambda^{\alpha}\lambda^{\beta}\left[D_{\alpha\dot{\alpha}}, D_{\beta\dot{\beta}}\right] &= \left[\lambda^{\alpha} D_{\alpha\dot{\alpha}}, \lambda^{\beta} D_{\beta\dot{\beta}}\right] \\
&= \left[h^{-1}\bar{D}_{\dot{\alpha}}h, h^{-1}\bar{D}_{\dot{\beta}}h\right] \\
&= h^{-1}\left[\bar{D}_{\dot{\alpha}}, \bar{D}_{\dot{\beta}}\right]h \\
&= 0.
\end{aligned} \tag{2.41}$$

for any λ_{α}. Thus decomposing the space-time curvature into irreducible components

$$\left[D_{\alpha\dot{\alpha}}, D_{\beta\dot{\beta}}\right] = \epsilon_{\alpha\beta} F_{\dot{\alpha}\dot{\beta}} + \epsilon_{\dot{\alpha}\dot{\beta}} F_{\alpha\beta} \tag{2.42}$$

we find that the self-dual curvature vanishes, $F_{\alpha\beta} = 0$. The non-vanishing component of the curvature is constructed via the non-linear Penrose transform

$$F_{\dot{\alpha}\dot{\beta}}(x) = \int_X \langle\lambda d\lambda\rangle\, h^{-1} \frac{\partial^2 a}{\partial\mu^{\dot{\alpha}}\partial\mu^{\dot{\beta}}}\, h \tag{2.43}$$

and obeys the equation of motion

$$D^{\alpha\dot{\alpha}} F_{\dot{\alpha}\dot{\beta}} = 0. \tag{2.44}$$

In summary, we have shown that given a holomorphic vector bundle $E \to U$ on an open region of twistor space $U \subset \mathbb{CP}^3$ that is topologically trivial $c_1(E) = 0$, we can construct a classical solutions of self-dual Yang-Mills theory on a corresponding region of space-time. It can be shown that this correspondence is in fact one-to-one [7].

2.2 $\mathcal{N} = 4$ Test Supersymmetric Yang–Mills Theory

2.2.1 Supertwistor Space

Let us now consider the $\mathcal{N} = 4$ supersymmetric extension of the above construction for self-dual Yang-Mills theory. Twistor space is now extended to the complex projective superspace $\mathbb{CP}^{3|4}$ by including four fermionic coordinates. Concretely, it is defined by the homogeneous coordinates

$$\mathcal{Z}^I = (\lambda_\alpha, \mu^{\dot\alpha}, \chi^a) \tag{2.45}$$

and the equivalence relation $\mathcal{Z}^I \sim r\,\mathcal{Z}^I$ for any $r \in \mathbb{C}^*$. The fermionic coordinates χ^a transform in the fundamental representation of an SU(4) R-symmetry. It will be important that supertwistor space is a Calabi-Yau supermanifold, meaning that it has a canonical top holomorphic form,

$$D^{3|4}\mathcal{Z} = \frac{1}{4!}\epsilon_{IJKL}Z^I dZ^J dZ^K dZ^L\, d^4\chi \tag{2.46}$$

a feature that is unique to maximal supersymmetry.

There is now a relationship between twistor space and the chiral superspace $\mathbb{CM}^{4|8}$ with coordinates $\left(x^{\alpha\dot\alpha}, \theta^{\alpha a}\right)$, where again the fermionic coordinates transform in the fundamental of SU(4). The correspondence is encapsulated by the extended incidence relations

$$\mu^{\dot\alpha} = i\, x^{\alpha\dot\alpha}\lambda_\alpha \qquad \chi^a = \theta^{\alpha a}\lambda_\alpha. \tag{2.47}$$

Given a point $\left(x^{\alpha\dot\alpha}, \theta^{\alpha a}\right)$ in chiral superspace, the incidence relations define a holomorphically embedded complex projective line in twistor space, $X \cong \mathbb{CP}^1$. Furthermore, as for the bosonic correspondence, two points in chiral superspace are null separated, that is

$$(x_1 - x_2)^2 = 0 \quad (x_1 - x_2)\cdot(\theta_1 - \theta_2) = 0 \tag{2.48}$$

if and only if the corresponding twistor lines intersect. Thus, the complex structure of supertwistor space determines and is determined by the superconformal structure of chiral superspace.

The construction of α-planes requires some additional explanation compared to the purely bosonic case. Given a point in supertwistor space $\mathcal{Z}^I$, the general solution of the incidence relations is now

$$x^{\alpha\dot\alpha} = x_0^{\alpha\dot\alpha} + \lambda^\alpha\tilde\lambda^{\dot\alpha} \qquad \theta^{\alpha a} = \theta_0^{\alpha a} + \lambda^\alpha\eta^a \tag{2.49}$$

where $(\tilde\lambda^{\dot\alpha}, \eta^a)$ are parameters labelling the solutions. The parameters span a superplane $\mathbb{C}^{2|4}$ inside chiral superspace $\mathbb{CM}^{4|8}$. This plane is completely null, meaning that all tangent vectors are orthogonal and null, in the sense of Eq. (2.48). These are the α-planes.

Twistor space is an important tool for studying superconformal field theories because it a carries a particularly natural action of the superconformal group. For maximal $\mathcal{N} = 4$ supersymmetry we have an infinitessimal action of the general linear supergroup GL(4|4, $\mathbb{C}$) by

$$J^I{}_J = Z^I \frac{\partial}{\partial Z^J}. \tag{2.50}$$

In order to generate the superconformal group SL(4|4, $\mathbb{C}$) we must ensure the generators have vanishing supertrace by removing a component proportional to $(-1)^I \delta^I{}_J\, b$ where $b = (-1)^K J^K{}_K$. In the following, we almost always consider homogeneous functions of the coordinates Z^I of weight zero, in which case $h = 0$ and we recover the action of PSL(4|4, $\mathbb{C}$) on supertwistor space $\mathbb{CP}^{3|4}$. In terms of spinor components, the generators (2.50) are

$$\begin{aligned} P_{\alpha\dot{\alpha}} &= \lambda_\alpha \frac{\partial}{\partial \mu^{\dot{\alpha}}} \qquad J_{\alpha\beta} = \frac{1}{2}\left(\lambda_\alpha \frac{\partial}{\partial \lambda^\beta} + \lambda_\beta \frac{\partial}{\partial \lambda^\alpha}\right) \qquad J_{\dot{\alpha}\dot{\beta}} = \frac{1}{2}\left(\mu_{\dot{\alpha}} \frac{\partial}{\partial \mu^{\dot{\beta}}} + \mu_{\dot{\beta}} \frac{\partial}{\partial \mu^{\dot{\alpha}}}\right) \\ K^{\alpha\dot{\alpha}} &= \mu^{\dot{\alpha}} \frac{\partial}{\partial \lambda_\alpha} \qquad D = \frac{1}{2}\left(\lambda_\alpha \frac{\partial}{\partial \lambda_\alpha} - \mu^{\dot{\alpha}} \frac{\partial}{\partial \mu^{\dot{\alpha}}}\right) \qquad R^a_{\ b} = \chi^a \frac{\partial}{\partial \chi^b}, \end{aligned} \tag{2.51}$$

and

$$\begin{aligned} Q_{\alpha a} &= \lambda_\alpha \frac{\partial}{\partial \chi^a} \qquad \widetilde{Q}^a_{\dot{\alpha}} = \chi^a \frac{\partial}{\partial \mu^{\dot{\alpha}}} \\ S^{\alpha a} &= \chi^a \frac{\partial}{\partial \lambda_\alpha} \qquad \widetilde{S}^{\dot{\alpha}}_a = \mu^{\dot{\alpha}} \frac{\partial}{\partial \chi^a}. \end{aligned} \tag{2.52}$$

Finally, the real form relevent for lorentzian signature, PSU(2, 2|4), consists of those transformations that preserve the same pseudo-hermitian metric (2.18) as in the bosonic case.

2.2.2 The Self-Dual Theory

Let us now construct the self-dual sector of $\mathcal{N} = 4$ supersymmetric Yang-Mills theory from the twistor space perspective. The argument is an extension of the purely bosonic case presented above. We introduce a smooth complex vector bundle of rank N

$$E \longrightarrow U \subset \mathbb{CP}^{3|4}, \tag{2.53}$$

with vanishing first Chern class $c_1(E) = 0$ and an almost complex structure $\bar{D} = \bar{\partial} + \mathcal{A}$. In the supersymmetric case, it is important that the partial connection is a one-form with components in the bosonic directions of supertwistor space only, that is we have $\mathcal{A} = \mathcal{A}_{\bar{I}}\, d\bar{Z}^{\bar{I}}$.

In perturbation theory, the bundle is again automatically holomorphic and holomorphically trivial once pulled back to any line $X \subset U$. Hence we can construct a holomorphic frame $\mathrm{H}(X, \lambda, \bar{\lambda})$ that obeys

$$\bar{D}_\lambda \mathrm{H} = 0 \tag{2.54}$$

and depends smoothly on the line X, or equivalently, on the coordinates $(x^{\alpha\dot{\alpha}}, \theta^{\alpha a})$. Furthermore, we can find N linearly independent globally holomorphic sections, which are unique up to constant GL(N) transformations. The space of such holomorphic sections $\Gamma(X, E|_X) \cong \mathbb{C}^{\mathrm{N}}$ now form the fibres of a complex vector bundle on a region of chiral superspace $\mathbb{CM}^{4|8}$, with complexified gauge group GL(N).

In order to construct a superconnection on the chiral superspace, we must impose further conditions on the almost complex structure. Pulling back to the spin bundle, the components of $\bar{D}$ are

$$\begin{aligned}\bar{D}_\lambda &= \bar{\partial}_\lambda + \mathcal{A}_\lambda \\ \bar{D}_{\dot{\alpha}} &= \lambda^\alpha \partial_{\alpha\dot{\alpha}} + \mathcal{A}_{\dot{\alpha}} \\ \bar{D}_a &= \lambda^\alpha \partial_{\alpha a}\end{aligned} \tag{2.55}$$

where $\mathcal{A}_\lambda$ and $\mathcal{A}_{\dot{\alpha}}$ are defined exactly as in the bosonic case. Again, the curvature $[\bar{D}_\lambda, \bar{D}_\lambda] = 0$ for dimensional reasons and, to construct a space-time superconnection, we must impose the further conditions

$$\begin{aligned}\left[\bar{D}_\lambda, \bar{D}_{\dot{\alpha}}\right] &= 0 \\ \left[\bar{D}_\lambda, \bar{D}_a\right] &= 0.\end{aligned} \tag{2.56}$$

These conditions imply that the combinations $\mathrm{H}^{-1} D_{\dot{\alpha}}\, \mathrm{H}$ and $\mathrm{H}^{-1} \bar{D}_a \mathrm{H}$ are holomorphic in λ_α with weight one, and hence

$$\begin{aligned}\mathrm{H}^{-1} \bar{D}_{\dot{\alpha}}\, \mathrm{H} &= \lambda^\alpha\, \mathcal{A}_{\alpha\dot{\alpha}}(x, \theta) \\ \mathrm{H}^{-1} \bar{D}_a\, \mathrm{H} &= \lambda^\alpha\, \mathcal{A}_{\alpha a}(x, \theta)\end{aligned} \tag{2.57}$$

where the space-time fields $(\mathcal{A}_{\alpha\dot{\alpha}}, \mathcal{A}_{\alpha a})$ transform as a superconnection under super gauge transformations that preserve the holomorphic frame. They allow the construction of space-time super-covariant derivatives $(\nabla_{\alpha\dot{\alpha}}, \nabla_{\alpha a})$.

An important difference compared to the bosonic case is that the conditions impose space-time equations of motion for the scalars and fermions. The reason is that the superconnection we constructed automatically obeys the integrability constraints

$$\begin{aligned}\lambda^\alpha \lambda^\beta [\, \nabla_{\alpha\dot{\alpha}}\,, \nabla_{\beta b}\,] &= 0 \\ \lambda^\alpha \lambda^\beta \{\, \nabla_{a\alpha}\,, \nabla_{\beta b}\, \} &= 0.\end{aligned} \tag{2.58}$$

These integrability constraints are absent in bosonic theory. In the supersymmetric theory, they imply some equations of motion for the scalar and fermion partners of the gluon. Therefore, the space-time superconnection cannot be constructed completely off-shell, causing subtle technical problems for the quantum theory when a regulator is introduced—the consequences of this are further discussed in Chap. 6. This is in fact a well-known problem with standard superspace approaches to space-time gauge theories with extended supersymmetry and can be remedied by harmonic superspace methods [8, 9].

Finally, we impose the remaining condition

$$\left[\bar{D}_{\dot{\alpha}}, \bar{D}_{\dot{\beta}}\right] = 0 \tag{2.59}$$

meaning that the twistor bundle is now holomorphic on the whole of U. This implies that the curvature of the space-time superconnection $\mathcal{A}_{\alpha\dot{\alpha}}$ obeys

$$\lambda^{\alpha}\lambda^{\beta}\left[\nabla_{\alpha\dot{\alpha}}, \nabla_{\beta\dot{\beta}}\right] = 0. \tag{2.60}$$

or equivalently that the corresponding supercurvature is self-dual, $\mathcal{F}_{\alpha\beta} = 0$. These are the remaining equations of motion of the self-dual theory. Thus, combining Eqs. (2.58) and (2.60), the equations of motion of the self-dual theory are equivalent to the statement that the space-time superconnection is flat when projected onto α-planes.

Since the above construction is rather abstract, let us now expand in components fields and derive the corresponding space-time equations of motion. We begin by expanding the partial connection in powers of the fermionic components

$$\begin{aligned}\mathcal{A}(Z,\chi) = a(Z,\bar{Z}) &+ \chi^a\tilde{\gamma}_a(Z,\bar{Z}) + \frac{1}{2!}\chi^a\chi^b\phi_{ab}(Z,\bar{Z})\\ &+ \frac{1}{3!}\epsilon_{abcd}\chi^a\chi^b\chi^c\gamma^d(Z,\bar{Z}) + \chi^1\chi^2\chi^3\chi^4 g(Z,\bar{Z}).\end{aligned} \tag{2.61}$$

The condition that the bundle be holomorphic

$$\bar{\partial}\mathcal{A} + \mathcal{A}\wedge\mathcal{A} = 0, \tag{2.62}$$

implies the following equations for the component fields

$$\begin{aligned}&\bar{\partial} + a\wedge a = 0\\ &\bar{D}\gamma_a = 0\\ &\bar{D}\phi^{ab} - \phi^a\wedge\phi^b = 0\\ &\bar{D}\,\bar{\gamma}^a - \frac{1}{2}\varepsilon^{abcd}\left(\gamma_b\wedge\phi_{cd} + \phi_{cd}\wedge\gamma_b\right) = 0\\ &\bar{D}g + \gamma_a\wedge\bar{\gamma}^a + \frac{1}{4}\varepsilon^{abcd}\phi_{ab}\wedge\phi_{cd},\end{aligned} \tag{2.63}$$

where $\bar{D} = \bar{\partial} + a$ is the bosonic part of the twistor superfield $\mathcal{A}$.

The space-time component fields are now constructed from the components of the partial connection $\mathcal{A}$ by the Penrose transform. The key ingredient in the transform is the bosonic part of the holomorphic frame $h(x, \lambda, \bar{\lambda})$ obtained by discarding higher order terms in the fermionic expansion

$$H(x, \theta, \lambda, \bar{\lambda}) = h(x, \lambda, \bar{\lambda}) + \cdots . \tag{2.64}$$

The space-time component fields are then constructed by the following Penrose transforms

$$\begin{aligned}
F_{\dot{\alpha}\dot{\beta}}(x) &= \int_X \langle\lambda \mathrm{d}\lambda\rangle\, h^{-1}(x, \lambda)\, \frac{\partial^2 a}{\partial\mu^{\dot{\alpha}}\mu^{\dot{\beta}}}(\lambda)\, h(x, \lambda) \\
\tilde{\psi}_{\dot{\alpha}a}(x) &= \int_X \langle\lambda \mathrm{d}\lambda\rangle\, h^{-1}(x, \lambda)\, \frac{\partial \tilde{\gamma}_a}{\partial\mu^{\dot{\alpha}}}(\lambda)\, h(x, \lambda) \\
\tilde{\phi}_{ab}(x) &= \int_X \langle\lambda \mathrm{d}\lambda\rangle\, h^{-1}(x, \lambda)\, \tilde{\phi}_{ab}(\lambda)\, h(x, \lambda) \\
\psi_{\alpha}{}^{a}(x) &= \int_X \langle\lambda \mathrm{d}\lambda\rangle\, h^{-1}(x, \lambda)\, \lambda_\alpha\, \gamma^a(\lambda)\, h(x, \lambda) \\
G_{\alpha\beta}(x) &= \int_X \langle\lambda \mathrm{d}\lambda\rangle\, h^{-1}(x, \lambda)\, \lambda_\alpha\, \lambda_\beta\, g(\lambda)\, h(x, \lambda).
\end{aligned} \tag{2.65}$$

and transform in the adjoint representation of the complexified gauge group $\mathrm{GL}(\mathrm{N}, \mathbb{C})$ under bosonic gauge transformations. The auxilliary field $G_{\alpha\beta}$ is non-dynamical. The remaining dynamical fields are the gauge field $A_{\alpha\dot{\alpha}}$, complex Weyl fermions ψ^a_α in the fundamental representation **4**, and scalars ϕ_{ab} in the antisymmetric tensor **6** of the R-symmetry group $\mathrm{SU}(4)_R$. These fields form the $\mathcal{N} = 4$ supermultiplet.

The component form of the holomorphic condition (2.63) and the definition of the bosonic space-time connection, $h^{-1} D_{\dot{\alpha}} h = \lambda^\alpha A_{\alpha\dot{\alpha}}$, then imply the following equations of motion for the space-time fields

$$\begin{aligned}
F_{\alpha\beta} &= 0 \\
D^{\alpha\dot{\alpha}} \bar{\psi}_{\dot{\alpha}a} &= 0 \\
\Box\phi_{ab} - \left\{\phi^{\dot{\alpha}}_a, \phi_{\dot{\alpha}b}\right\} &= 0 \\
D^{\alpha\dot{\alpha}} \psi^a_\alpha + \mathrm{i}\left[\phi^{ab}, \phi^{\dot{\alpha}}_b\right] &= 0 \\
D^{\alpha\dot{\alpha}} G_{\alpha\beta} + \mathrm{i}\left\{\psi^a_\beta, \bar{\psi}^{\dot{\alpha}}_a\right\} + D_{\beta\dot{\alpha}}\left[\phi^{ab}, \phi_{ab}\right] &= 0\,.
\end{aligned} \tag{2.66}$$

These are precisely the field equations for self-dual $\mathcal{N} = 4$ supersymmetric Yang-Mills theory, obtained from the space-time action [10, 11],

$$\mathcal{L}_1 = \mathrm{tr}\Big\{ -\frac{1}{4} G^{\alpha\beta} F_{\alpha\beta} + i\bar{\psi}_{\dot{\alpha}a} D^{\dot{\alpha}\alpha} \psi^a_\alpha + \frac{1}{4} D_{\alpha\dot{\alpha}}\phi^{ab} D^{\alpha\dot{\alpha}}\phi_{ab} + \bar{\psi}_{\dot{\alpha}a}\big[\phi^{ab}, \bar{\psi}^{\dot{\alpha}}_b\big]\Big\}\,. \tag{2.67}$$

2.2.3 The Complete Theory

We formulate the complete $\mathcal{N}=4$ supersymmetric Yang-Mills theory as an expansion around the self-dual sector of the theory, which had a particularly simple construction in twistor space. The additional interaction lagrangian is

$$\mathcal{L}_2 = g^2 \mathrm{tr}\Big\{ -\frac{1}{8} G^{\alpha\beta} G_{\alpha\beta} - \psi^{\alpha a}[\bar{\phi}_{ab}, \psi^b_\alpha] + \frac{1}{8}[\phi^{ab}, \phi^{cd}][\bar{\phi}_{ab}, \bar{\phi}_{cd}]\Big\} \tag{2.68}$$

and leads to the equations of motion

$$\begin{aligned}
&F_{\alpha\beta} = g^2\, G_{\alpha\beta} \\
&D^{\alpha\dot{\alpha}} \bar{\psi}_{\dot{\alpha} a} = g^2\, \mathrm{i}\, [\phi_{ab}, \psi^{\alpha a}] \\
&\Box\phi_{ab} - \left\{\phi^{\dot{\alpha}}_a, \phi_{\dot{\alpha} b}\right\} = \frac{g^2}{2}\epsilon_{abcd}\left\{\psi^{\alpha c}, \psi^d_\alpha\right\} + g^2\left[\phi_{ac}, \left[\phi_{bd}, \phi^{cd}\right]\right] \\
&D^{\alpha\dot{\alpha}}\psi^a_\alpha + \mathrm{i}\left[\phi^{ab}, \bar{\psi}^{\dot{\alpha}}_b\right] = 0 \\
&D^{\alpha\dot{\alpha}} G_{\alpha\beta} + \mathrm{i}\left\{\psi^a_\beta, \phi^{\dot{\alpha}}_a\right\} + D_{\beta\dot{\alpha}}\left[\phi^{ab}, \phi_{ab}\right] = 0.
\end{aligned} \tag{2.69}$$

The corrections to the self-dual sector all appear proportional to the squared coupling g^2 on the right-hand side.

Let us concentrate on the relationship between the self-dual component of the curvature $F_{\alpha\beta}$ and the auxilliary field $G_{\alpha\beta}$. The components of the lagrangian involving the auxilliary field are

$$-\frac{1}{4}\mathrm{tr}\Big\{G^{\alpha\beta} F_{\alpha\beta} - \frac{1}{2} g^2 G^{\alpha\beta} G_{\alpha\beta}\Big\} \tag{2.70}$$

and the equations of motion $F_{\alpha\beta} = g^2 G_{\alpha\beta}$ reduce to the self-dual condition as $g^2 \to 0$. In this manner we are expanding around the self-dual sector. On the other hand, integrating out the auxilliary field we obtain the complete space-time action for maximally supersymmetric Yang-Mills theory,

$$\begin{aligned}
\mathcal{L} = \mathrm{tr}\Big\{ &-\frac{1}{4g^2} F^{\alpha\beta} F_{\alpha\beta} + i\bar{\psi}_{\dot{\alpha} a} D^{\dot{\alpha}\alpha}\psi^a_\alpha + \frac{1}{4} D_{\alpha\dot{\alpha}}\phi^{ab} D^{\alpha\dot{\alpha}}\phi_{ab} \\
&+ \bar{\psi}_{\dot{\alpha} a}[\phi^{ab}, \bar{\psi}^{\dot{\alpha}}_b] - g^2\psi^{\alpha a}[\bar{\phi}_{ab}, \psi^b_\alpha] + \frac{g^2}{8}[\phi^{ab}, \phi^{cd}][\bar{\phi}_{ab}, \bar{\phi}_{cd}]\Big\}
\end{aligned} \tag{2.71}$$

This is indeed the unique, local, gauge invariant space-time action that is invariant under superconformal symmetries PSU(2,2|4).

2.2.4 The Twistor Action

We begin with the self-dual sector of $\mathcal{N} = 4$ supersymmetric Yang-Mills theory. In the previous section, we explained that classical solutions are obtained from holomorphic vector bundles on supertwistor space, which are characterised by the vanishing of the (0, 2)-component of the curvature

$$\bar{\partial}\mathcal{A} + \mathcal{A} \wedge \mathcal{A} = 0. \tag{2.72}$$

Modulo complex gauge transformations. Thus holomorphic vector bundles on supertwistor space can be obtained as the stationary points of holomorphic Chern-Simons theory on twistor space,

$$S_1 = \frac{1}{2\pi} \int D^{3|4}\mathcal{Z} \wedge \mathrm{Tr}\left(\mathcal{A} \wedge \bar{\partial}\mathcal{A} + \frac{2}{3}\mathcal{A} \wedge \mathcal{A} \wedge \mathcal{A} \right). \tag{2.73}$$

It is the unique gauge-invariant local action that depends only on the complex structure and the holomorphic volume form $D^{3|4}\mathcal{Z}$. The holomorphic Chern-Simons action was introduced in [12] and shown to reduce, in an appropriate gauge, to the self-dual action (2.67) in reference [13].

In order to expand the complete theory around the self-dual sector, we add further interaction terms to the holomorphic Chern-Simons theory. The interaction involves the logarithm of the determinant of the complex structure $(\bar{\partial} + \mathcal{A})|_X$ integrated over all lines inside null supertwistor space [13]. In the following, we will write this interaction term as

$$S_2 = g^2 \int \mathrm{d}^{4|8}X \;\log\det \bar{D}\big|_X , \tag{2.74}$$

although strictly speaking, one should subract the contribution $\log\det \bar{\partial}\big|_X$ from the trivial background connection. The determinant is actually a section of the determinant line bundle over the space of partial connections on the bundle $E|_X$ and picks up anomalous contributions under gauge transformations. However, the logarithm ensures that these contributions are additive and annihilated by the fermionic integral [13].

The determinant is not convenient for explicit calculations. However, the correction can be expanded in a power series in the twistor partial connection $\mathcal{A}$ leading to an infinite series of interaction terms (this expansion is derived in Chap. 5),

$$g^2 \sum_{n=2}^{\infty} \int \mathrm{d}^{4|8}X \int_{X^n} \frac{\mathrm{d}\rho_n \cdots \mathrm{d}\rho_1}{(\rho_1 - \rho_n) \cdots (\rho_2 - \rho_1)(\rho_1 - \rho_n)} \,\mathrm{Tr}\, \mathcal{A}(\rho_n) \cdots \mathcal{A}(\rho_1). \tag{2.75}$$

where each coordinate $\{\rho_1, \ldots, \rho_n\}$ is a complex coordinate on the line X. This expansion allows perturbative calculations to be performed straightforwardly in twistor space.

References

1. T. Adamo, M. Bullimore, L. Mason, D. Skinner, Scattering amplitudes and wilson loops in twistor space. J. Phys.A **A44**, 454008 (2011). http://arxiv.org/abs/1104.2890
2. R. Penrose, W. Rindler, *Spinors and Space-Time. Spinor and Twistor Methods in Space-Time Geometry*. (Cambridge University Press, Cambridge, 1987)
3. S. Huggett, K. Tod, *An Introduction to Twistor Theory*. (Cambridge University Press, Cambridge, 1994)
4. R. Penrose, Solutions of the zero-rest-mass equations. J. Math. Phys. **10**, 38–39 (1969)
5. R. Penrose, On the twistor description of massless fields. In *Proceedings Complex Manifold Techniques In Theoretical Physics*, 1978, ed. by Lawrence, pp. 55–91
6. R. Penrose, M.A. MacCallum, Twistor theory: an approach to the quantization of fields and space-time. Phys. Rep. **6**, 241–316 (1972)
7. R. Ward, On selfdual gauge fields. Phys. Lett. **A61**, 81–82 (1977)
8. A. Galperin, E. Ivanov, S. Kalitsyn, V. Ogievetsky, E. Sokatchev, Unconstrained N=2 matter, Yang-Mills and supergravity theories in harmonic superspace. Class. Quant. Grav. **1**, 469–498 (1984)
9. A. Galperin, E. Ivanov, V. Ogievetsky, E. Sokatchev, *Harmonic Superspace*. (Cambridge University Press, 2007), p. 306
10. G. Chalmers, W. Siegel, The selfdual sector of QCD amplitudes. Phys. Rev. **D54**, 7628–7633 (1996). http://arxiv.org/abs/hep-th/9606061
11. W. Siegel, N = 2, N = 4 string theory is selfdual N = 4 Yang-Mills theory. Phys. Rev. **D46**, 3235–3238 (1992)
12. E. Witten, Perturbative gauge theory as a string theory in twistor space, Commun. Math. Phys. **252**, 189–258 (2004). http://arxiv.org/abs/hep-th/0312171
13. R. Boels, L. Mason, D. Skinner, Supersymmetric Gauge Theories in Twistor Space, JHEP **0702**, 014 (2007). http://arxiv.org/abs/hep-th/0604040

Chapter 3
Amplitudes and MHV Diagrams

Scattering amplitudes are fundamental and remarkably rich observables in quantum field theory. Scattering amplitudes in gauge theories are often much simpler than one expects from a typical Feynman diagram expansion. The first hint of the stunning simplifications that can occur came with a discovery of Parke and Taylor [1]. Using the spinor–helicity formalism, they found a beautiful expression for the tree-level scattering amplitude of six gluons in a simple helicity configuration

$$A(1^-, 2^-, 3^+, 4^+, 5^+, 6^+) = \frac{\langle 12\rangle^4}{\langle 12\rangle\langle 23\rangle \ldots \langle 61\rangle}. \tag{3.1}$$

This expression is equivalent to summing over two hundred rather complicated Feynman diagrams. The above expression is just the beginning of the remarkable simplicity and hidden structures that have been discovered in gauge theory scattering amplitudes.

Much progress has been made in computational efficiency and structural understanding with the introduction of purely on-shell methods. For tree-level amplitudes, there is the connected prescription of twistor-string theory [2, 3], on-shell recursion relations [4, 5], the CSW/MHV diagram expansion [6, 7] and unitarity based methods, see for example [8–10]. The above techniques bring to the fore different properties of scattering amplitudes, but they are all characterized by constructing amplitudes while remaining completely on-shell, and are consequently vastly more efficient than Feynman diagram methods.

When looking for new theoretical structures and computational techniques, it is often useful to first examine the simplest theory and then extend the lessons learned to more complicated examples. In this spirit, we here consider scattering amplitudes in maximally supersymmetric $\mathcal{N} = 4$ Yang-Mills theory. This particular theory has provided significant inspiration in developing the aforementioned on-shell techniques and has provided many important theoretical insights. This quantum field theory is also of theoretical interest because of the remarkable duality with type IIB string theory in $\mathrm{AdS}_5 \times \mathrm{S}^5$ [11–13] and this connection has lead to surprising connections with other areas of mathematical physics.

M. R. Bullimore, *Scattering Amplitudes and Wilson Loops in Twistor Space*, Springer Theses, DOI: 10.1007/978-3-319-00909-4_3,

In this chapter, we first present a brief review of the colour decomposition of amplitudes and the powerful spinor-helicity formalism. We then introduce a remarkable hidden symmetry of partial amplitudes in $\mathcal{N} = 4$ supersymmetric gauge theory in the planar limit, dual superconformal symmetry, and explain how to construct invariants under this hidden symmetry using momentum twistor variables. Finally we reformulate the MHV diagram formalism in momentum twistor space and find dramatic simplifications, allowed manifestly dual superconformally invariant expressions to be generated for all tree-level amplitudes and loop integrands with remarkable ease. This chapter provides the springboard for subsequent developments in Chaps. 4 and 5.

3.1 Colour Decomposition

Let us now make some preliminary comments on colour decomposition and the planar limit. Much more information and explanations can be found, for example, in the reviews [14, 15].

In the following, we consider gauge theories with coupling constant g^2 and gauge group SU(N) with hermitian generators denoted by T^a where $a = 1, \ldots, N^2 - 1$. Then for an l-loop correction to n-gluon scattering amplitudes we have the colour decomposition [16]

$$(g^2N)^l \sum_{\sigma \in S_n/\mathbb{Z}_n} \mathrm{Tr}\left[T^{a_{\sigma(1)}} \ldots T^{a_{\sigma(n)}}\right] A^{(l)}\left(\sigma(1^{h_1}), \ldots, \sigma(n^{h_n}), N\right) + \ldots \tag{3.2}$$

where h_n denote the helicities of the incoming particles and the $+\ldots$ stands for terms involving multiple traces that are supressed by additional powers of $1/N$. The partial amplitudes $A^{(l)}\left(1^{h_1}, \ldots, n^{h_n}, N\right)$ are colour ordered meaning they only recieve contributions from Feynman diagrams with a particular cyclic ordering of the gluons. Consequently, they have a restricted set of physical singularities involving only sums of adjacent momenta, for example $p_1 + p_2 + \cdots + p_j$.

Furthermore, we are concerned here only with scattering amplitudes in the planar limit, where $g^2 \to 0$, $N \to \infty$ and the 't Hooft coupling $\lambda \equiv g^2N$ is held fixed. In the planar limit, the multiple trace terms are then supressed and the partial amplitudes are independent of N,

$$A^{(l)}\left(1^{h_1}, \ldots, n^{h_n}, N\right) \to A^{(l)}\left(1^{h_1}, \ldots, n^{h_n}\right) + \mathcal{O}(1/N)\,. \tag{3.3}$$

The above structure can be derived by a careful analysis of the Feynman diagrams or perhaps more elegantly from the structure of open string theory scattering amplitudes [16]. From now on we consider exclusively the colour-ordered partial amplitudes in the planar limit.

3.2 Kinematics

In any Poincare invariant quantum field theory in four dimensions, massless scattering states are labelled by their four-momentum $p^{\alpha\dot{\alpha}}$ and helicity h. The massless on-shell condition $p^2 = 0$ implies that

$$p^{\alpha\dot{\alpha}} = \lambda^{\alpha}\tilde{\lambda}^{\dot{\alpha}} \,. \tag{3.4}$$

This decomposition is invariant under little group transformations

$$\lambda^{\alpha} \to t\lambda^{\alpha} \qquad \tilde{\lambda}^{\dot{\alpha}} \to t^{-1}\tilde{\lambda}^{\dot{\alpha}} \tag{3.5}$$

where $t \in \mathbb{C}^*$. This is dimply a description of the complex light-cone as a complex cone over $\mathbb{CP}^1 \times \mathbb{CP}^1$. For real four-momentum in Minkowski signature, $\tilde{\lambda}^{\dot{\alpha}} = \pm\bar{\lambda}^{\dot{\alpha}}$ and the parameter $t = e^{i\theta/2}$ is a phase. The wave-functions of massless states are then required to transform

$$\psi(\lambda, \tilde{\lambda}) \to t^{-2h}\psi(\lambda, \tilde{\lambda}) \tag{3.6}$$

under little group transformations, where h is the helicity.

Now consider the momentum and polarization wavefunctions of the on-shell states. For scalars there is no polarization wavefunction and the momentum space wavefunction is plane wave

$$\phi(x) = \exp\left(ix^{\alpha\dot{\alpha}}\lambda_{\alpha}\tilde{\lambda}_{\dot{\alpha}}\right). \tag{3.7}$$

For negative helicity fermions, we are looking for plane wave solutions to the equation $i\,\partial^{\alpha\dot{\alpha}}\psi_{\alpha} = 0$. The unique solution, up to normalisation, is

$$\psi_{\alpha}(x) = \lambda_{\alpha} \exp\left(ix^{\alpha\dot{\alpha}}\lambda_{\alpha}\tilde{\lambda}_{\dot{\alpha}}\right) \tag{3.8}$$

and similarly for positive helicity fermions

$$\tilde{\psi}_{\dot{\alpha}}(x) = \tilde{\lambda}_{\alpha} \exp\left(ix^{\alpha\dot{\alpha}}\lambda_{\alpha}\tilde{\lambda}_{\dot{\alpha}}\right). \tag{3.9}$$

The wavefunctions clearly transform with the correct weight under little group transformations.

For massless gluons of positive and negative helicity, we are looking for plane wave solutions of the form

$$A_{\alpha\dot{\alpha}}(x) = \epsilon_{\alpha\dot{\alpha}} \exp\left(ix^{\alpha\dot{\alpha}}\lambda_{\alpha}\tilde{\lambda}_{\dot{\alpha}}\right) \tag{3.10}$$

where the polarization vector obeys the Lorentz gauge constraint $\epsilon_{\alpha\dot{\alpha}}\lambda^{\alpha}\tilde{\lambda}^{\dot{\alpha}} = 0$, representing the decoupling of longitudinal modes. In addition, those related by gauge transformations $\epsilon^{\alpha\dot{\alpha}} \to \epsilon^{\alpha\dot{\alpha}} + \alpha\,\lambda^{\alpha}\tilde{\lambda}^{\dot{\alpha}}$ represent the same physical on-shell state, so the polarisation vectors describe two degrees of freedom. The polarization vectors are constructed by choosing auxilliary spinors ζ_α and $\tilde{\zeta}_{\dot\alpha}$, not proportional to λ_α and $\tilde{\lambda}_{\dot\alpha}$ respectively, and forming the combinations

$$\epsilon^-_{\alpha\dot\alpha} = \frac{\lambda_\alpha \tilde{\zeta}_{\dot\alpha}}{[\tilde{\lambda}, \tilde{\zeta}]} \qquad \epsilon^+_{\alpha\dot\alpha} = \frac{\zeta_\alpha \tilde{\lambda}_{\dot\alpha}}{\langle \zeta, \lambda\rangle}. \tag{3.11}$$

Changes in the auxilliary spinor $\tilde{\zeta}_{\dot\alpha} \to \alpha\, \tilde{\lambda}_{\dot\alpha} + \beta\, \tilde{\zeta}_{\dot\alpha}$ correspond to gauge transformations of $\epsilon^-_{\alpha\dot\alpha}$, and similarly for $\epsilon^+_{\alpha\dot\alpha}$.

The linearised curvature for the positive and negative helicity solutions are easily shown to be self-dual and anti-self-dual respectively. The wavefunctions

$$F_{\dot\alpha\dot\beta} = \tilde{\lambda}_{\dot\alpha}\tilde{\lambda}_{\dot\beta} \exp\left(i x^{\alpha\dot\alpha}\lambda_\alpha\tilde{\lambda}_{\dot\alpha}\right) \qquad F_{\alpha\beta} = \lambda_\alpha\lambda_\beta \exp\left(i x^{\alpha\dot\alpha}\lambda_\alpha\tilde{\lambda}_{\dot\alpha}\right) \tag{3.12}$$

are independent of the auxilliary spinors. Again, the wavefunctions clearly transform with the correct weight under little group transformations.

3.3 Symmetries

3.3.1 Superconformal Symmetry

We now collect the wavefunctions of on-shell states in the $\mathcal{N} = 4$ supermultiplet

$$\begin{aligned} F_{\dot\alpha\dot\beta} &= \tilde{\lambda}_{\dot\alpha}\tilde{\lambda}_{\dot\beta}\, g^+(\lambda, \tilde{\lambda}) \\ \tilde{\psi}_{\dot\alpha a} &= \tilde{\lambda}_{\dot\alpha}\, \tilde{\Gamma}_a(\lambda, \tilde{\lambda}) \\ \phi_{ab} &= \phi_{ab}(\lambda, \tilde{\lambda}) \\ \psi^a_\alpha &= \lambda_\alpha\, \Gamma^a(\lambda, \tilde{\lambda}) \\ F_{\alpha\beta} &= \lambda_\alpha\lambda_\beta\, g^-(\lambda, \tilde{\lambda}). \end{aligned} \tag{3.13}$$

The momentum space wavefunctions of the on-shell states in the $\mathcal{N} = 4$ supermultiplet can be combined into a single on-shell superfield

$$\begin{aligned} \Phi(\lambda, \tilde{\lambda}, \eta) = {} & g^+(\lambda, \tilde{\lambda}) + \eta^a \tilde{\Gamma}_a(\lambda, \tilde{\lambda}) + \frac{1}{2!}\eta^a\eta^b\phi_{ab}(\lambda, \tilde{\lambda}) \\ & + \frac{1}{3!}\epsilon_{abcd}\eta^a\eta^b\eta^c\Gamma^d(\lambda, \tilde{\lambda}) + \frac{1}{4!}\epsilon_{abcd}\eta^a\eta^b\eta^c\eta^d g^-(\lambda, \tilde{\lambda}.). \end{aligned} \tag{3.14}$$

where the fermionic parameters η^a transform in the fundamental of $SU(4)_R$ and transform in the same way as $\tilde{\lambda}$ under little group rescalings. Hence the on-shell superfield transforms with weight -2. Scattering amplitudes are combined into superamplitudes $\mathcal{A}(\lambda_i, \tilde{\lambda}_i, \eta_i)$ which describe simultaneously the scattering of all on-shell states in the theory.

The superconformal symmetry $\mathfrak{psu}(2,2|4)$ of the lagrangian is represented by the on-shell superfields and places important constraints on the superamplitudes. The bosonic subalgebra $\mathfrak{su}(2,2) \oplus \mathfrak{su}(4)$ contains the conformal algebra $\mathfrak{su}(2,2) \cong \mathfrak{so}(2,4)$ and the R-symmetry $\mathfrak{su}(4)$. The conformal algebra is generated by the Poincare generators

$$p^{\alpha\dot{\alpha}} = \sum_i \lambda_i^\alpha \tilde{\lambda}_i^{\dot{\alpha}} \tag{3.15}$$

$$m_{\alpha\beta} = \sum_i \left(\lambda_{i\alpha} \frac{\partial}{\partial \lambda_i^\beta} + \lambda_{i\beta} \frac{\partial}{\partial \lambda_i^\alpha} \right) \qquad \tilde{m}_{\dot{\alpha}\dot{\beta}} = \sum_i \left(\lambda_{i\alpha} \frac{\partial}{\partial \lambda_i^\beta} + \lambda_{i\beta} \frac{\partial}{\partial \lambda_i^\alpha} \right), \tag{3.16}$$

and the conformal dilations and boosts

$$d = \sum_i \lambda_i^\alpha \frac{\partial}{\partial \lambda_i^\alpha} \qquad k_{\alpha\dot{\alpha}} = \sum_i \frac{\partial}{\partial \lambda_{i\alpha} \partial \tilde{\lambda}_{i\dot{\alpha}}}. \tag{3.17}$$

The R-symmetry algebra $\mathfrak{su}(4)$ represented by

$$R^a{}_b = \sum_i \left(\eta_i^a \frac{\partial}{\partial \eta_i^b} - \frac{1}{4} \delta_b^a \, \eta_i^c \frac{\partial}{\partial \eta_i^c} \right). \tag{3.18}$$

In addition, we have the Poincaré and conformal supercharges

$$\begin{aligned} q^{\alpha a} &= \sum_i \lambda_i^\alpha \eta_i^a \qquad & \bar{q}_a^{\dot{\alpha}} &= \sum_i \tilde{\lambda}_i^{\dot{\alpha}} \frac{\partial}{\partial \eta_i^a} \\ s_{\alpha a} &= \sum_i \frac{\partial^2}{\partial \lambda_i^\alpha \partial \eta_i^a} \qquad & \bar{s}_{\dot{\alpha}}^a &= \sum_i \eta_i^a \frac{\partial}{\partial \tilde{\lambda}_i^{\dot{\alpha}}}. \end{aligned} \tag{3.19}$$

Invariance under the translation generator $P^{\alpha\dot{\alpha}}$ implies that amplitudes are proportional to the momentum conserving delta-function. Similarly, invariance under the supersymmetry generator $Q_\alpha{}^a$ requires the presence of a fermionic supermomentum conserving delta-function. In addition, the correct weight under little group rescalings may be obtained pulling out the Parke-Taylor denomenator, as follows

$$\mathcal{M}_n(\lambda_j, \tilde{\lambda}_j, \eta_j) = \frac{\delta^{(0|8)}\left(\sum_{j=1}^n \lambda_j \eta_j\right)}{\langle \lambda_1 \lambda_2 \rangle \langle \lambda_2 \lambda_3 \rangle \cdots \langle \lambda_n \lambda_1 \rangle} \mathcal{A}_n(\lambda_j, \tilde{\lambda}_j, \eta_j). \tag{3.20}$$

The remaining amplitude $\mathcal{A}_n$ is invariant under $\mathrm{SU}(4)_\mathrm{R}$ transformations of the η^a coordinates and has zero weight under rescalings. Since $\mathrm{SU}(4)_\mathrm{R}$ invariant combinations are formed from multiples of four η^a coordinates we have an expansion

$$\mathcal{A}_n(\lambda_j, \tilde{\lambda}_j, \eta_j) = 1 + \sum_{k=1}^{n-4} \mathcal{A}_{n,k}(\lambda_j, \tilde{\lambda}_j, \eta_j) \tag{3.21}$$

where the $\mathcal{A}_{n,k}$ are homogeneous polynomials of order $\mathcal{O}(\eta^{4k})$. The superamplitude $\mathcal{A}_{n,k}$ is called the $\mathrm{N^k}$MHV remainder function and in particular contains the scattering amplitude of $k+2$ negative helicity gluons and $n-k-2$ positive helicity gluons. The remaining ordinary superconformal generators place further constraints on the superamplitudes, although we do not consider them here.

3.3.2 Dual Superconformal Symmetry

For planar superamplitudes, the colour ordering allows the momentum and supermomentum conservation conditions

$$\sum_{j=1}^{n} \lambda_j^{\alpha} \tilde{\lambda}_j^{\dot{\alpha}} = 0 \qquad \sum_{j=1}^{n} \lambda_j^{\alpha} \eta_j^{a} = 0 \tag{3.22}$$

to be solved automatically by introducing region coordinates $\{x_j^{\alpha\dot{\alpha}}, \theta_j^{\alpha a}\}$ whose cyclic separations are the momenta

$$x_{j+1}^{\alpha\dot{\alpha}} - x_j^{\alpha\dot{\alpha}} = \lambda_j^{\alpha} \tilde{\lambda}_j^{\dot{\alpha}} \qquad \theta_{j+1}^{\alpha a} - \theta_j^{\alpha a} = \lambda_j^{\alpha} \eta_j^{a} \tag{3.23}$$

Since the momenta are invariant under a common superspace translation, the region coordinates live in complex affine superspace $\mathbb{CM}^{4|8}$, called the dual superspace. Thus the kinematics is equivalently determined by ordered set of points in $\mathbb{CM}^{4|8}$ that are sequentially null separated (Fig. 3.1).

The kinematics of superamplitudes are equivalently determined by choosing an ordered set of supertwistors $\{\mathcal{Z}_1, \ldots, \mathcal{Z}_n\}$, constrained only by the condition

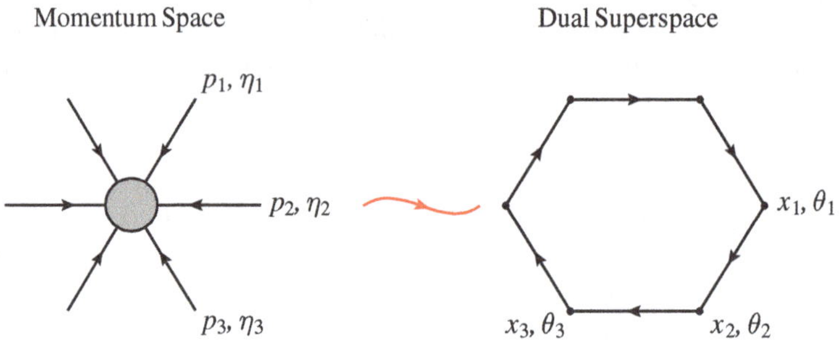

Fig. 3.1 The kinematics of superamplitudes are encoded in a null polygon in dual superspace

$$\langle i-1, i\rangle \equiv I_{IJ}\mathcal{Z}^I_{i-1}\mathcal{Z}^J_i \neq 0, \tag{3.24}$$

where $i = 1, 2, \ldots, n$, which ensures we are away from a collinear limit [17]. The momentum twistors determine a system of complex projective lines $\{X_1, \ldots, X_n\}$ which intersect sequentially at the momentum twistors, $\mathcal{Z}_j = X_{j-1} \cap X_j$. This in turn determines a sytem of null separated points in chiral superspace, whose coordinates are

$$x_j^{\alpha\dot{\alpha}} = i\,\frac{\lambda^\alpha_{j-1}\mu^{\dot{\alpha}}_j - \lambda^\alpha_j\mu^{\dot{\alpha}}_{j-1}}{\langle j-1, j\rangle} \qquad \theta_j^{\alpha a} = \frac{\lambda^\alpha_{j-1}\chi^a_j - \lambda^\alpha_j\chi^a_{j-1}}{\langle j-1, j\rangle}. \tag{3.25}$$

Equivalently, the momentum twistors each determine an α-plane in chiral superspace, whose sequential intersections are the dual coordinates. The superamplitudes now become homogeneous functions of the momentum supertwistors.

$$\mathcal{A}_{n,k}(\mathcal{Z}_1, \ldots, \mathcal{Z}_n) \tag{3.26}$$

The important point about this construction is that the twistor data $\{\mathcal{Z}_1, \ldots, \mathcal{Z}_n\}$ is completely unconstrained. Thus momentum twistors completely solve the constraints of momentum conservation and the null conditions for on-shell supermomenta. This is the reason that momentum twistors are so useful.

Another important property of momentum twistors is that they manifest an additional superconformal symmetry of planar superamplitudes in $\mathcal{N} = 4$ supersymmetric Yang-Mills theory. This new symmetry is simply the superconformal symmetry of the dual superspace and is known as 'dual' superconformal symmetry. Dual conformal symmetry was originally observed at strong coupling via the amplitude / Wilson loop correspondence [18] and in perturbative computations at weak coupling [19–22]. The dual conformal symmetry of tree-level planar superamplitudes was extended to dual superconformal symmetry in [23] and its existance has been understood as a consequence of T-duality of IIB string theory on the $\mathrm{AdS}_5 \times \mathrm{S}^5$ background [24, 25]. In the authors opinion, the most compelling understanding of dual superconformal symmetry comes from the superamplitude / super Wilson loop correspondence, which is the topic of Chaps. 5 and 6.

Furthermore, it has been shown that the standard superconformal symmetry of the lagrangian and dual superconformal symmetry together generate a Yangian symmetry $\mathcal{Y}(\mathfrak{psu}(2,2|4))$ of tree-level planar superamplitudes [26]. In the 'T-dual' representation of [27], the level-zero generators are the dual superconformal generators acting locally on the incoming particles[1]

$$J^{(0)I}{}_J = \sum_i \mathcal{Z}^I_i \frac{\partial}{\partial \mathcal{Z}^J_i}\,. \tag{3.27}$$

[1] In Eqs. (3.27) and (3.28) it is understood that the the supertrace component of the generators are removed—see discussion in Sect. 2.2.1.

In momentum twistors space, the original superconformal symmetries $p_{\alpha\dot{\alpha}}$ and q^a_α have been trivialised, while the remaining superconformal generators form a subset of the level-one Yangian generators in the bilocal representation of [28, 29],

$$J^{(1)I}{}_J = \sum_{i<j}\left((-1)^K \mathcal{Z}_i^I \frac{\partial}{\partial \mathcal{Z}_i^K} \mathcal{Z}_j^K \frac{\partial}{\partial \mathcal{Z}_j^J} - (i \leftrightarrow j)\right). \tag{3.28}$$

It has been shown that the above representation of the level-zero and level-one generators obey the required Serre relations [26] and therefore generate a representation of the complete Yangian, meaning that all generators $J^{(2)}, J^{(3)}, \ldots$ then also annihilate planar tree-level superamplitudes.

Tree-level superamplitudes are themselves constructed from linear combinations of so-called leading singularities of loop amplitudes, whose many intricate relationships are beautifully encoded in the grassmannian formulae of [30–32]. It is now understood that all leading singularities are individually Yangian invariant and that the grassmannian formulae encode all such invariants [33–35]. However, many of the Yangian generators are broken in the full loop amplitudes, but the anomalies have been understood and can provide powerful constraints. For example, the breaking of dual special conformal and dilatation generators, K and D, is well-known to control the infrared structure of loop corrections—see for example [19, 20]. Some of the dual supersymmetries, $\bar{Q}$ and S, are broken even for the infrared-finite remainder and the anomalies have recently been understood in [36, 37]. The breaking of dual superconformal symmetry is discussed in Chap. 6.

3.3.3 Dual Superconformal Invariants

Let us now consider the construction of dual superconformal invariants. We can immeadiately write down bosonic dual conformal invariants using the SL(4,$\mathbb{C}$) invariant skew tensor,

$$\langle i, j, k, l\rangle \equiv \epsilon_{IJKL} Z_i^I Z_j^J Z_k^K Z_l^L\,. \tag{3.29}$$

The fundamental dual superconformally invariant object was first constructed in dual superspace coordinates in [23, 38, 39]. Here we use the simpler momentum twistor construction and define the object [32]:

$$\begin{aligned}[i, j, k, l, m] &= \int_{\mathbb{CP}^4} \frac{\mathrm{D}^4 c}{c_1 c_2 c_3 c_4 c_5}\, \delta^{4|4}\left(c_1 \mathcal{Z}_i + c_2 \mathcal{Z}_j + c_3 \mathcal{Z}_k + c_4 \mathcal{Z}_l + c_5 \mathcal{Z}_m\right) \\ &= \frac{\delta^{0|4}\left(\langle i, j, k, l\rangle \chi_m + \text{cyclic}\right)}{\langle i, j, k, l\rangle \langle j, k, l, m\rangle \langle k, l, m, i\rangle \langle l, m, i, j\rangle \langle m, i, j, k\rangle}\end{aligned} \tag{3.30}$$

and is completely antisymmetric under interchange of the momentum twistor arguments. Any five bosonic momentum twistors are automatically linearly dependent and the dual superconformal bracket is simply a fermionic delta-function ensuring the same is true for five supertwistors.

In fact, we will eventually encounter a whose family of superconformally invariant delta-functions, which impose various linear dependences on their momentum twistor arguments [32, 40]. The first member of the series,

$$\delta^{3|4}(\mathcal{Z}_1, \mathcal{Z}_2) = \int \frac{\mathrm{d}c}{c} \delta^{4|4}(\mathcal{Z}_1 + c\mathcal{Z}_2) \tag{3.31}$$

has support when $\mathcal{Z}_1 = r\mathcal{Z}_2$ for some non-zero complex number r. This means that the supertwistors are forced to coincide at the same point in $\mathbb{CP}^{3|4}$. Integrating over one additional parameter we find

$$\delta^{2|4}(\mathcal{Z}_1, \mathcal{Z}_2, \mathcal{Z}_3) := \int \frac{\mathrm{d}c_1}{c_1} \frac{\mathrm{d}c_2}{c_2} \delta^{4|4}(\mathcal{Z}_1 + c_1\mathcal{Z}_2 + c_3\mathcal{Z}_3)\,, \tag{3.32}$$

which has support when the arguments are collinear in $\mathbb{CP}^{3|4}$. This delta-function plays an important role in the propagator for holomorphic Chern-Simons theory in axial gauge [41]. Continuing, we find the coplanarity delta-function $\delta^{1|4}(\mathcal{Z}_1, \mathcal{Z}_2, \mathcal{Z}_3, \mathcal{Z}_4)$ and finally the dual superconformal bracket $[\mathcal{Z}_1, \mathcal{Z}_2, \mathcal{Z}_3, \mathcal{Z}_4, \mathcal{Z}_5]$. All of these objects are completely anti-symmetric in their arguments.

3.3.4 The Loop Integrand

As mentioned above, dual superconformal symmetry is broken in loop corrections to superamplitudes by infrared divergences and the need to introduce a regularization scheme. In planar theories, the infrared divergences can be temporarily avoided by 'removing' the loop integration and introducing the notion of the four-dimensional loop integrand [33]. This is a manifestly finite and Yangian invariant object with much remarkable structure in its own right. It is also the object that appears naturally in the mometum twistor space approach to the amplitude / Wilson loop correspondence to be discussed in Chap. 5.

Let us now describe what is meant by 'four-dimensional integrand' following [33]. Given a lagrangian description of the theory at hand, scattering amplitudes can be computed by summing the appropriate Feynman diagrams. For an l-loop correction, each Feynman diagram involves an integral over l loop momenta, say $q_1, \ldots, q_l$, which might be divergent and require regularization. However, in general there is no canonical way to combine the Feynman diagrams under the same loop integral because there is no canonical origin for momentum space. For example, one might translate the loop momentum in one diagram by $q \to q + p_1$ where p_1 is the momentum of an external particle.

For planar superamplitudes, however, this obstruction can be overcome. We have already seen that momentum conservation can be solved by introducing dual

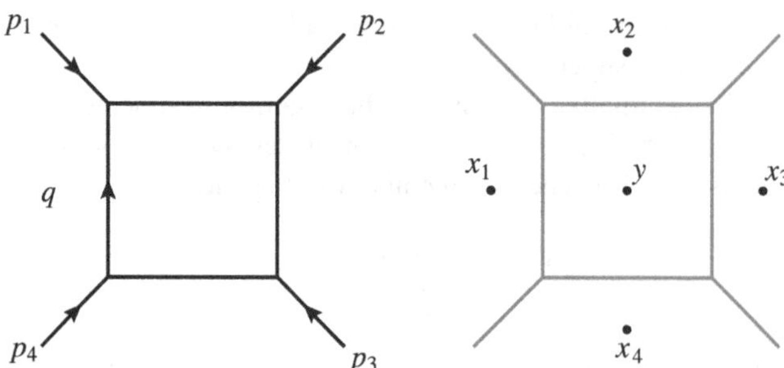

Fig. 3.2 A simple one-loop example of assigning dual coordinates to internal regions. The loop momentum is reconstructed as $q = x_1 - y$

coordinates $x_1, \ldots, x_n$ from which the momenta can be constructed via $p_i = x_{i+1} - x_i$. Similarly, dual coordinates $y_1, \ldots, y_l$ can now be assigned to all internal regions of Feynman diagrams, and the momenta flowing through the legs constructed from differences of the dual coordinates. This is illustrated for a simple box diagram in Fig. 3.2. The assignment of coordinates to internal regions is unique up to permutations of the internal dual coordinates. Thus, provided we symmetrise over such assignments, all Feynman diagrams can be combined under a single loop integral involving the coordinates $y_1, \ldots, y_l$.

The Feynman diagram expression for the l-loop correction to a planar scattering amplitude is then

$$A(p_1, \ldots, p_n) = \int \mathrm{d}^4 y_1 \ldots \mathrm{d}^4 y_n \; I(p_1, \ldots, p_n; y_1, \ldots, y_\ell) \,. \tag{3.33}$$

The four-dimensional integrand I in this expression is completely well-defined. However, the amplitude on the right-hand side is certainly not well-defined since the integrals might diverge and require regularization. Such potential infrared and ultraviolet divergences are physically meaningful and important, so the four-dimensional integrand is limited in its scope. Indeed, the four-dimensional integrand does not quite uniquely determine the regularised integrand in dimensional or the Coulomb branch regularization of [42] and, in certain circumstances, this can be very important [34].

In planar $\mathcal{N} = 4$ supersymmetric Yang-Mills theory, we consider a supersymmetric extension of the four-dimensional integrand where for each loop we assign a superspace coordinate $(y^{\alpha\dot{\alpha}}, \theta^{\alpha a})$ or equivalently a complex line Y is supertwistor space. The (unregulated) superamplitude is then

$$\mathcal{A}^{(\ell)}_{n,k}(\mathcal{Z}_1, \ldots, \mathcal{Z}_n) = \int \mathrm{d}^{4|8} Y_1 \ldots \mathrm{d}^{4|8} Y_n \;\; \mathcal{I}^{(\ell)}_{n,k}(\mathcal{Z}_1, \ldots, \mathcal{Z}_n; Y_1, \ldots, Y_\ell) \,. \tag{3.34}$$

In order to obtain a uniform treatment, we will often introduce momentum twistors $\{A_j, B_j\}$ to describe the internal loop coorinates, $Y_j = A_j \cap B_j$. The four-dimensional integrands are then finite rational functions of the momentum twistors. When the generators are extended to act on the internal momentum twistors, the loop integrands are invariant under dual superconformal and Yangian symmetry [33]. In Chaps. 3–5, we deal exclusively with the supersymmetric four-dimensional loop integrand.

3.4 MHV Diagram Formalism

The MHV diagram formalism is a Feynman diagram-like expansion where the vertices are MHV amplitudes, continued off-shell by introducing an auxiliary reference spinor, and the propagators are scalar Feynman propagators [6]. It offers substantial simplifications on the standard Feynman rules and leads to compact expressions for planar superamplitudes and integrands.

Originally formulated for tree-level scattering amplitudes of gluons, the MHV diagram formalism has since been extended to theories including scalars and fermions [43–46] and has been used to compute one-loop corrections to amplitudes in supersymmetric gauge theories [47–49]. The lagrangian origin of the MHV diagram formalism has also been uncovered both from the space-time perspective [50, 51] and from the twistor action [52].

In this chapter, we develop a supersymmetric version of the MHV diagram formalism for planar superamplitudes and extend it to four-dimensional loop integrands. In momentum twistor space, the diagrammatic rules are simple and uniform [53, 54]. The vertices now contribute '1' and propagators are associated with dual superconformal brackets whose arguments are assigned by a simple geometric rule. This provides an efficient method to generate manifestly dual superconformal and cyclic expressions for tree-level superamplitudes and loop integrands.

3.4.1 *Momentum Space*

In the maximally supersymmetric version of the MHV diagram formalism the vertices are MHV superamplitudes

$$\mathcal{A}_{\mathrm{MHV}}(1,\ldots,n) = \frac{\delta^{0|8}\Big(\sum_{j=1}^{n} \lambda_j \eta_j\Big)}{\langle 1,2\rangle \ldots \langle n,1\rangle} \tag{3.35}$$

and the propagators are massless scalar Feynman propagators

$$\frac{1}{p^2 + i\epsilon}. \tag{3.36}$$

Since internal propagators have some off-shell momentum p flowing in them, we must have some prescription for constructing a holomorphic spinor $|p\rangle$ associated to this momentum. Here, we choose an auxiliary spinor $|\zeta]$ and then associate

$$|p\rangle^{\alpha} = p^{\alpha\dot{\alpha}}|\zeta]_{\dot{\alpha}} \tag{3.37}$$

with the off-shell momentum $p^{\alpha\dot{\alpha}}$. Finally, for each internal propagator we perform a fermionic integration $\mathrm{d}^4\eta$ which implements the sum over on-shell states propagating in the channel.

Individual diagrams will depend on the auxiliary spinor $\zeta^{\dot{\alpha}}$ and this dependence must cancel out in the sum over diagrams. This assertion can be proved by deriving the formalism from a recursive argument and we shall do so in Chap. 4. We also note that the dependence on $\zeta^{\dot{\alpha}}$ leads to spurious singularities in individual diagrams, for example

$$\frac{1}{\langle 1|(p_2 + p_3)|\zeta]}, \tag{3.38}$$

and that all spurious singularities involve $\zeta^{\dot{\alpha}}$ in this way. Hence, independence of the reference spinor implies that all such unphysical singularities must indeed cancel in the sum over all diagrams.

3.4.2 Momentum Twistor Space

In this section we reformulate the supersymmetric MHV diagram formalism in momentum twistor space, and find that vertices now contribute '1' and propagators are associated with dual superconformal brackets [, , , ,]. We proceed by concrete example before explaining the general rule for assigning dual superconformal brackets to propagators.

NMHV Tree Diagrams

The simplest diagrams have two vertices connected by a propagator, as illustrated in Fig. 3.3. Such diagrams have Grassmann degree four and hence contribute to the tree-level NMHV superamplitude.

The diagram has the momentum space expression

$$\int \mathrm{d}^4\eta\, \mathcal{A}_{\mathrm{MHV}}(i, \ldots, j-1, \{\lambda, \eta\}) \frac{1}{(x_i - x_j)^2} \mathcal{A}_{\mathrm{MHV}}(\{\lambda, \eta\}, j, \ldots, i-1) \tag{3.39}$$

where $|\lambda\rangle = (x_i - x_j)\,|\zeta]$ is the left-handed spinor assigned to the off-shell momentum in the propagator. The first step is to extract the overall tree-level MHV superamplitude, leaving the expression

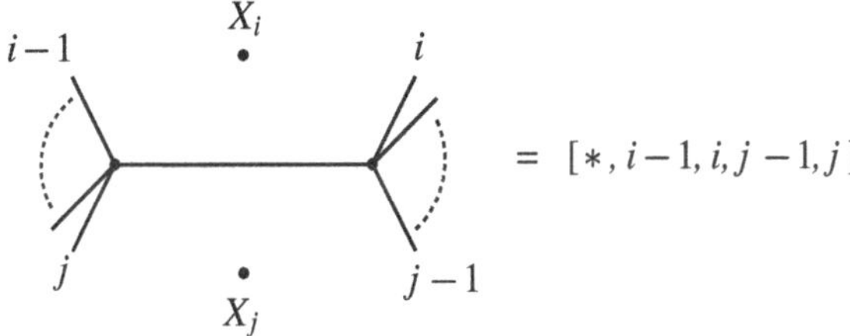

Fig. 3.3 A generic diagram contributing to the NMHV tree-level superamplitude

$$\frac{\langle i-1, i\rangle\langle j-1, j\rangle}{(x_i-x_j)^2\langle i-1, \lambda\rangle\langle\lambda, i\rangle\langle j-1, \lambda\rangle\langle\lambda, j\rangle} \times \int d^4\eta\, \delta^{0|8}\big(\theta_i - \theta_j - |\lambda\rangle\eta\big)\,. \tag{3.40}$$

The fermionic integration is straightforward to perform, resulting in the expression

$$\frac{\langle i-1, i\rangle\langle j-1, j\rangle\, \delta^{0|4}\big([\zeta|x_{ij}|\theta_{ji}\rangle\big)}{x_{ij}^2\, [\zeta|x_{ij}|i-1\rangle\, [\zeta|x_{ij}|i\rangle\, [\zeta|x_{ij}|j-1\rangle\, [\zeta|x_{ij}|j\rangle}\,. \tag{3.41}$$

In order to translate to momentum twistor space we introduce an auxiliary momentum twistor $\mathcal{Z}_* = (0, \zeta^{\dot{\alpha}}, 0)$ constructed from the auxiliary spinor. The factors in the denominator of the expression (3.41) may now be written in terms of momentum twistors as, for example,

$$[\zeta|x_{ji}|i\rangle = \frac{\langle *, j-1, j, i\rangle}{\langle j-1, j\rangle}\,, \qquad x_{ij}^2 = \frac{\langle i-1, i, j-1, j\rangle}{\langle i-1, i\rangle\langle j-1, j\rangle}\,, \tag{3.42}$$

while the argument of the remaining fermionic delta-function becomes

$$\begin{aligned} [i|x_{ij}|\theta_{ji}\rangle &= \langle *, i-1, i, \big[j-1\rangle, \chi_j\big] + \langle *, j-1, j, [i-1\rangle, \chi_i] \\ &= \langle *, i-1, i, j-1\rangle\chi_j + \text{cyclic}\,, \end{aligned} \tag{3.43}$$

where in the second line we recall that the reference twistor $\mathcal{Z}_*$ has vanishing fermionic components. The expression in Eq. (3.41) is now just the dual superconformal invariant

$$\begin{aligned} &[*, i-1, i, j-1, j] \\ &= \frac{\delta^{0|4}(\langle *, i-1, i, j-1\rangle\chi_j + \text{cyclic})}{\langle i-1, i, j-1, j\rangle\langle i, j-1, j, *\rangle\langle j-1, j, *, i-1\rangle\langle j, *, i-1, i\rangle\langle *, i-1, i, j-1\rangle} \end{aligned} \tag{3.44}$$

and the complete expression for the NMHV tree-level superamplitude is therefore

$$\mathcal{A}_{n,1}^{(0)} = \sum_{i<j} [*, i-1, i, j-1, j]\,. \tag{3.45}$$

From the known dual superconformal symmetry [23] and independence of the momentum space formalism on the reference spinor $\zeta^{\dot{\alpha}}$, the components of the reference twistor $\mathcal{Z}_*$ may be changed to arbitrary values by dual superconformal transformations. The independence of the reference twistor may be seen directly in this case by expanding each term using the linear identity

$$[i-1, i, j-1, j, *'] + [i, j-1, j, *', *] + [j-1, j, *', *, i-1] \\ + [j, *', *, i-1, i] + [*', *, i-1, i, j-1] + [*, i-1, i, j-1, j] = 0. \quad (3.46)$$

The terms depending on the original reference twistor $\mathcal{Z}_*$ cancel in pairs, leaving the same diagrammatic expansion with a new reference twistor $\mathcal{Z}_{*'}$.

Choosing the reference twistor to be the momentum twistor $\mathcal{Z}_n$ and recalling the complete anti-symmetry of dual superconformal brackets, we find the known BCFW expansion of the NMHV tree-level superamplitude [33]:

$$\mathcal{A}_{n,1}^{(0)} = \sum_{1<i<j<n} [n, i-1, i, j-1, j] \quad (3.47)$$

This expression is no longer manifestly cyclic invariant but displays manifest invariance under dual superconformal symmetry.

N^{2}MHV Tree

Now consider diagrams containing three vertices connected by two propagators. Such diagrams have grassmann degree eight and hence contribute to the tree-level N^{2}MHV superamplitude. The generic such diagrams are illustrated in Fig. 3.4 and in this notation the range is restricted to $j \geq i+2$ and $l \geq k+2$ since vertices with fewer than three legs vanish. For generic MHV diagrams with $l > i$ and $k > j$, there are no adjacent propagators connected to the central vertex and the calculation proceeds as two separate copies of our previous example, leading to

$$[\, * \, , i-1, i, j-1, j] [\, * \, , k-1, k, l-1, l]\, . \quad (3.48)$$

However, in boundary cases, $l = i$ or $j = k$, we must carefully translate the momentum space expression into momentum twistor space.

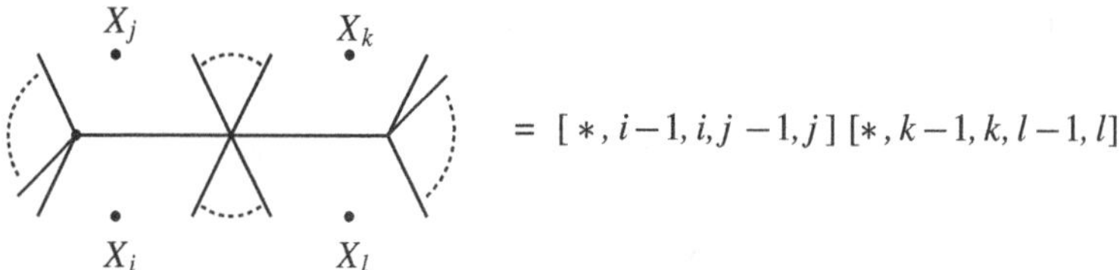

Fig. 3.4 A generic MHV diagram for the N^2 MHV tree-level superamplitude

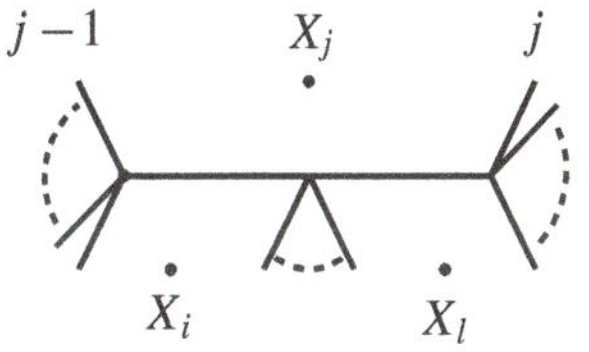

$j-1$ X_j j

X_i X_l

$$= [*, i-1, i, j-1, j'] [*, j-1, j, l-1, l]$$

Fig. 3.5 A boundary MHV diagram for the N^2 MHV tree-level superamplitude

Let us consider the diagram with $j = k$ as shown in Fig. 3.5. There are now two adjacent propagators connected to the central vertex. Pulling out an overall MHV superamplitude, the momentum space rules lead to the expression

$$\frac{\langle i-1\, i\rangle\langle j-1\, j\rangle\, \delta^{0|4}\Big([\zeta\, |x_{ij}|\theta_{ji}\rangle\Big)}{x_{ij}^2\, \langle i-1\, \lambda\rangle\, \langle \lambda\, i\rangle\, \langle j-1\, \lambda\rangle} \times \frac{1}{\langle \lambda\, \lambda'\rangle} \times \frac{\langle l-1\, l\rangle\, \delta^{0|4}\Big([\zeta\, |x_{jl}|\theta_{lj}\rangle\Big)}{x_{jl}^2\, \langle \lambda' j\rangle\, \langle l-1 \lambda'\rangle\, \langle \lambda' l\rangle}, \tag{3.49}$$

where

$$\lambda^{\alpha} = (x_i - x_j)^{\alpha\dot{\alpha}} \zeta_{\dot{\alpha}} \qquad \text{and} \qquad \lambda'^{\alpha} = (x_j - x_l)^{\alpha\dot{\alpha}} \zeta_{\dot{\alpha}} \tag{3.50}$$

are the holomorphic spinors associated to the propagators. The expression now contains the new angled bracket factor $\langle \lambda\lambda'\rangle$ due to the adjacent propagators. Translating this to momentum twistor space we find

$$\langle \lambda\lambda'\rangle = \frac{\langle *, i-1, i, [j-1\rangle\langle j], l-1, l, *\rangle}{\langle i-1, i\rangle\langle j-1, j\rangle\langle l-1, l\rangle}. \tag{3.51}$$

and hence the result

$$[*, i-1, i, j-1, j'] [*, j-1, j, l-1, l] \tag{3.52}$$

where the momentum twistor $\mathcal{Z}'_j = \langle *, l-1, l, [\mathcal{Z}_{j-1}\rangle\, \mathcal{Z}_j]$ is the intersection of the line $(j-1, j)$ and the plane $(*, l-1, l)$. This is a natural point in the twistor geometry of such boundary terms, which is illustrated in Fig. 3.6.

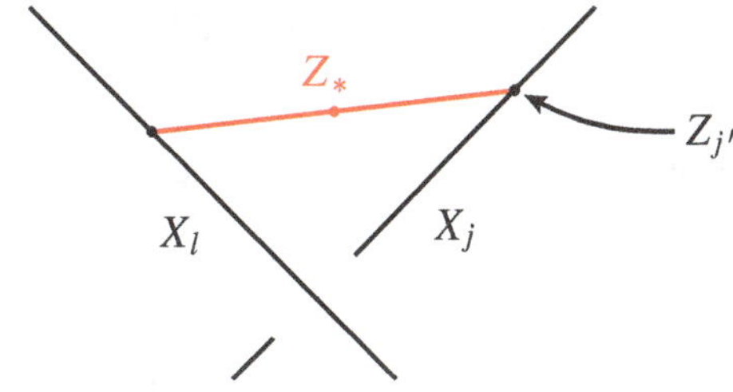

Fig. 3.6 The momentum twistor geometry associated with boundary contributions to the tree-level N^2 MHV superamplitude

The complete tree-level N^{2}MHV superamplitude may now be written

$$\mathcal{A}^{(0)}_{n,2} = \sum_{ij} [\,*\,, i-1, i, j-1, j'\,][\,*\,, k-1, k, l-1, l'\,] \tag{3.53}$$

where the sum is over the range $1 \leq i < j \leq k < l \leq i+n$ (understood modulo n) and where

$$\mathcal{Z}'_j = \begin{cases} \mathcal{Z}_j & \text{if } j < k \\ \langle *, l-1, l, [\mathcal{Z}_{j-1}\rangle\, \mathcal{Z}_j] & \text{if } j = k \end{cases} \tag{3.54}$$

and

$$\mathcal{Z}'_l = \begin{cases} \mathcal{Z}_l & \text{if } l < i \\ \langle *, j-1, j, \left[\mathcal{Z}_{l-1}\right\rangle \mathcal{Z}_l\right] & \text{if } l = i. \end{cases} \tag{3.55}$$

This expression agrees with the standard momentum space expression term-by-term if we choose the reference twistor $\mathcal{Z}_*$ to have spinor components $(0, \zeta^{\dot\alpha}, 0)$. However, the sum (3.53) is independent of $\zeta^{\dot\alpha}$ and dual superconformally invariant and is thus independent of the complete supertwistor $\mathcal{Z}_*$.

We conclude this subsection with two remarks. Firstly, the rule for assigning dual superconformal brackets requires that we choose an orientation for planar diagrams. With the opposite orientation the boundary diagram in Fig. 3.5 would be assigned an equivalent expression

$$[\,*\,, i-1, i, j-1, j][\,*\,, j-1', j, l-1, l]$$

where now $\mathcal{Z}'_{j-1} = \langle *, i-1, i, [\mathcal{Z}_{j-1}\rangle\, \mathcal{Z}_j]$. Secondly, the dual superconformal invariance of brackets involving shifted twistors follows only on the support of another unshifted bracket. For example, the bracket $[*, j-1', j, l-1, l]$ is dual conformally invariant, but only dual superconformally invariant on the support of the bracket $[*, i-1, i, j-1, j]$ which ensures that the points $\{\mathcal{Z}_*, \mathcal{Z}_{i-1}, \mathcal{Z}_i, \mathcal{Z}_{j-1}, \mathcal{Z}_j\}$ span a four-dimensional subspace of the full supertwistor space $\mathbb{CP}^{3|4}$.

MHV 1-loop

The MHV diagrams contributing to the one-loop MHV superamplitude are shown in Fig. 3.7.

In momentum space, once we pull out the overall momentum conserving delta-function, the diagram in Fig. 3.7 leads to the following expression

$$\int d^4\eta_i\, d^4\eta_j\, \frac{A_{\mathrm{MHV}}(\ell_i, i, \ldots, j-1, \ell_j) A_{\mathrm{MHV}}(\ell_j, j \ldots, i-1, \ell_i)}{(x-x_i)^2 (x-x_j)^2}, \tag{3.56}$$

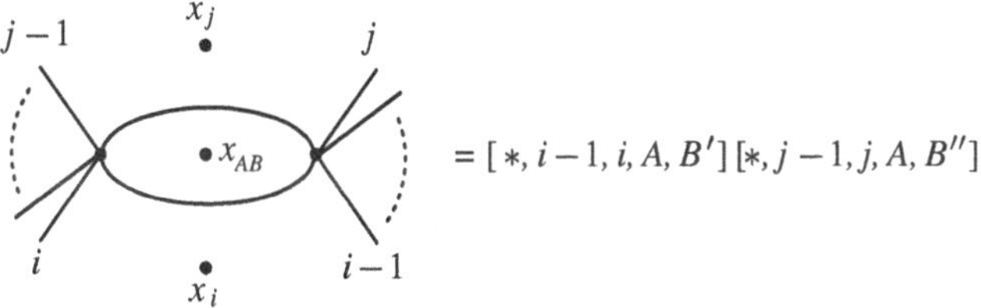

Fig. 3.7 Diagrams contributing to the one-loop MHV superamplitude

where

$$|\ell_i\rangle = (x - x_i)|\zeta] \quad \text{and} \quad |\ell_j\rangle = (x - x_j)|\zeta] \tag{3.57}$$

are the holomorphic spinors associated to the off-shell momenta in the lower and upper propagators, respectively. It is important that the propagators in Eq. (3.56) are Feynman propagators, including the correct $+i\epsilon$ prescription, in order to obtain the correct expressions when the integrals are performed in the dimensional regularization scheme [47, 55]. However, since we are focussing on the structure of four-dimensional integrands, the $+i\epsilon$ prescription is omitted in the following.

Pulling out an overall MHV tree-level superamplitude and performing the fermionic loop integrals, the remaining integrand is

$$\frac{1}{(x-x_i)^2(x-x_j)^2}\,\frac{\langle i-1, i\rangle\langle j-1, j\rangle\langle \ell_i\, \ell_j\rangle^2}{\langle i-1\, \ell_i\rangle\,\langle \ell_i\, i\rangle\,\langle j-1\, \ell_j\rangle\,\langle \ell_j\, j\rangle}\,. \tag{3.58}$$

To translate this expression to momentum twistor space, we again promote the reference spinor to a reference momentum twistor $\mathcal{Z}_* = (0, \zeta^{\dot{\alpha}}, 0)$. We then find the expression

$$\frac{1}{\langle a, b, i-1, i\rangle\,\langle a, b, j-1, j\rangle}\,\frac{\langle\, *, iv-1, i, [a\rangle, \langle b], j-1, j, \,*\,\rangle^2}{\langle a, b, i-1, *\rangle\,\langle a, b, i, *\rangle\,\langle a, b, j-1, *\rangle\,\langle a, b, j, *\rangle}\,. \tag{3.59}$$

where we have expressed the internal line $X = (a, b)$ as that passing through bosonic momentum twistors Z_a and Z_b. The numerator of this expression comes from $\langle \ell_i\, \ell_j\rangle^2$ and may be understood as the square of $\langle a, b, (*, i-1, i,) \cap (*, j-1, j)\rangle$. A simplification occurs in the boundary cases $j = i+1$ where the diagram contains a 3-point vertex with external particle i. The numerator in Eq. (3.58) simplifies using the relation

$$\begin{aligned}&\langle\, *, i-1, i, a\rangle\,\langle b, i, i+1, \,*\,\rangle + \langle\, *, i-1, i, b\rangle\,\langle a, i+1, i, \,*\,\rangle\\ &\quad + \langle\, *, i-1, i, i+1\rangle\,\langle a, b, i, \,*\,\rangle = 0\,.\end{aligned} \tag{3.60}$$

The resulting numerator factor $\langle a, b, i, *\rangle^2$ cancels two spurious propagators resulting in the expression

$$\frac{\langle i-1, i, i+1, *\rangle^2}{\langle a, b, i-1, i\rangle\langle a, b, i, i+1\rangle\langle a, b, i-1, *\rangle\langle a, b, *, i+1\rangle} \tag{3.61}$$

which is a box integrand with two spurious and two physical propagators.

While the dual conformal properties of the bosonic integrand (3.59) are clear, the dual superconformal properties are not. In other words, we would like to promote the bosonic twistors $Z_{a,b}$ to supertwistors $\mathcal{Z}_{a,b}$ and find an expression in terms of dual superconformal brackets. We note that the same numerator factor arises as the Jacobian when performing the following fermionic integrations

$$\int \mathrm{d}^4\chi_a \mathrm{d}^4\chi_b\, \delta^{0|4}\big(\langle *, i-1, i, a\rangle\chi_b' + \text{cyclic}\big)\delta^{0|4}\big(\langle *, j-1, j, a\rangle\chi_b'' + \text{cyclic}\big)$$
$$\langle *, i-1, i, a\rangle^4\langle *, j-1, j, a\rangle^4 \left(\langle *, i-1, i, [a\rangle, \langle b], j-1, j, *\rangle\right)^4 \tag{3.62}$$

where

$$\mathcal{Z}_b' = \langle *, j-1, j, \left[\mathcal{Z}_a\right\rangle \mathcal{Z}_b\big] \quad \text{and} \quad \mathcal{Z}_b'' = \langle *, i-1, i, \left[\mathcal{Z}_a\right\rangle \mathcal{Z}_b\big] \tag{3.63}$$

Hence the loop integrand in Eq. (3.58) may be written equivalently as

$$\int \mathrm{d}^4\chi_a \mathrm{d}^4\chi_b\, [*, i-1, i, a, b'][*, j-1, j, a, b'']. \tag{3.64}$$

and the fermionic loop integrations are absorbed into the supersymmetric loop integration measure leaving the product of two dual superconformal brackets.

Summing over all possible diagrams we find the that loop integrand of the MHV one-loop superamplitude may be written

$$\mathcal{I}_{n,0}^{(1)} = \sum_{i<j} [*, i-1, i, a, b'][*, j-1, j, a, b'']. \tag{3.65}$$

Note that the summation range $i < j$ here allows $i = j-1$, corresponding to a three-point MHV vertex on one end of the loop. The loop integrand is dual superconformally invariant, so again we are now free to choose any reference twistor. Incidentally, note again that choosing the reference twistor to be the external momentum twistor Z_n, the expansion (3.65) becomes

$$\mathcal{I}_{n,0}^{(1)} = \sum_{1<i<j<n} [n, i-1, i, a, b'][n, j-1, j, a, b''] \tag{3.66}$$

which reduces on performing the fermionic loop integrals to the loop integrand found in [33] using the BCFW recursion relation.

3.4.3 The General Case

It should be clear from the above examples that, in momentum twistor space, the vertices are assigned '1' and the propagators are assigned dual superconformal brackets [, , , ,]. It is now straightforward to give a concrete rule for assigning arguments for the dual superconformal brackets that treats internal and external regions in a uniform manner [53].

Consider a general propagator bounding regions U and V. Suppose further that the propagator is connected to vertices that have adjacent regions W and Y respectively in the anti-clockwise orientation, as illustrated in Fig. 3.8. The corresponding lines in momentum twistor space are

$$U = (U_1, U_2) \quad \text{and} \quad V = (V_1, V_2) \tag{3.67}$$

and similarly for the lines W and Y. We emphasize that these regions may be external $(\mathcal{Z}_{j-1}, \mathcal{Z}_j)$ or internal $(\mathcal{Z}_a, \mathcal{Z}_b)$. The rule for assigning dual superconformal brackets is blind to this distinction.

This propagator is assign the following dual superconformal bracket

$$[*, U_1, \widetilde{U}_2, V_1, \widetilde{V}_2] \tag{3.68}$$

where the arguments

$$\widetilde{U}_2 = (U_1, U_2) \cap (W_1, W_2, *) \tag{3.69}$$

and

$$\widetilde{V}_2 = (V_1, V_2) \cap (Y_1, Y_2, *) \tag{3.70}$$

are natural intersection points in the momentum twistor geometry. Notice that this rule depends on a choice of orientation around each vertex.

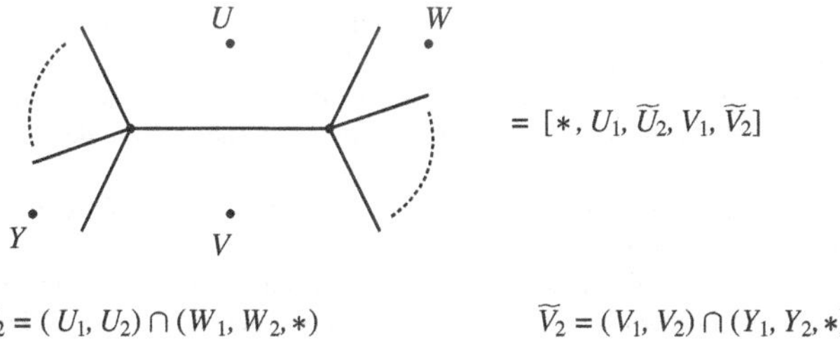

Fig. 3.8 General assignment of dual superconformal brackets to propagators

In the next chapter we will derive and prove the supersymmetric MHV formalism for planar tree-level superamplitudes and loop integrands.

References

1. S.J. Parke, T. Taylor, An amplitude for *n* gluon scattering. Phys. Rev. Lett. **56**, 2459 (1986)
2. E. Witten, Perturbative gauge theory as a string theory in twistor space. Commun. Math. Phys. **252**, 189–258 (2004, hep-th/0312171)
3. R. Roiban, M. Spradlin, A. Volovich, On the tree level S matrix of Yang-Mills theory. Phys. Rev. **D70**, 026009 (2004, hep-th/0403190)
4. R. Britto, F. Cachazo, B. Feng, New recursion relations for tree amplitudes of gluons. Nucl. Phys. **B715**, 499–522 (2005, hep-th/0412308)
5. R. Britto, F. Cachazo, B. Feng, E. Witten, Direct proof of tree-level recursion relation in Yang-Mills theory. Phys. Rev. Lett. **94**, 181602 (2005, hep-th/0501052)
6. F. Cachazo, P. Svrcek, E. Witten, MHV vertices and tree amplitudes in gauge theory. JHEP **0409**, 006 (2004, hep-th/0403047)
7. K. Risager, A direct proof of the CSW rules. JHEP **0512**, 003 (2005, hep-th/0508206)
8. Z. Bern, L.J. Dixon, D.C. Dunbar, D.A. Kosower, One loop n point gauge theory amplitudes, unitarity and collinear limits. Nucl. Phys. **B425**, 217–260 (1994, hep-ph/9403226)
9. Z. Bern, L.J. Dixon, D.C. Dunbar, D.A. Kosower, Fusing gauge theory tree amplitudes into loop amplitudes. Nucl. Phys. **B435**, 59–101 (1995, hep-ph/9409265)
10. C. Berger, Z. Bern, L. Dixon, F. Febres Cordero, D. Forde, et al., An automated implementation of on-shell methods for one-loop amplitudes. Phys. Rev. **D78**, 036003 (2008, arXiv:0803.4180)
11. J.M. Maldacena, The large N limit of superconformal field theories and supergravity. Adv. Theor. Math. Phys. **2**, 231–252 (1998, hep-th/9711200)
12. E. Witten, Anti-de Sitter space and holography. Adv. Theor. Math. Phys. **2**, 253–291 (1998, hep-th/9802150)
13. S. Gubser, I.R. Klebanov, A.M. Polyakov, Gauge theory correlators from noncritical string theory. Phys. Lett. **B428**, 105–114 (1998, hep-th/9802109)
14. L.J. Dixon, Calculating scattering amplitudes efficiently, hep-ph/9601359
15. L.J. Dixon, Scattering amplitudes: the most perfect microscopic structures in the universe. J. Phys.A **A44**, 454001 (2011, arXiv:1105.0771) * Temporary entry*
16. Z. Bern, D.A. Kosower, Color decomposition of one loop amplitudes in gauge theories. Nucl. Phys. **B362**, 389–448 (1991)
17. A. Hodges, Eliminating spurious poles from gauge-theoretic amplitudes, arXiv:0905.1473
18. L.F. Alday, J.M. Maldacena, Gluon scattering amplitudes at strong coupling. JHEP **0706**, 064 (2007, arXiv:0705.0303)
19. G. Korchemsky, J. Drummond, E. Sokatchev, Conformal properties of four-gluon planar amplitudes and Wilson loops. Nucl. Phys. **B795**, 385–408 (2008, arXiv:0707.0243)
20. J. Drummond, J. Henn, G. Korchemsky, E. Sokatchev, Conformal ward identities for Wilson loops and a test of the duality with gluon amplitudes. Nucl. Phys. **B826**, 337–364 (2010, arXiv:0712.1223)
21. J. Drummond, J. Henn, V. Smirnov, E. Sokatchev, Magic identities for conformal four-point integrals. JHEP **0701**, 064 (2007, hep-th/0607160)
22. Z. Bern, M. Czakon, L.J. Dixon, D.A. Kosower, V.A. Smirnov, The four-Loop planar amplitude and cusp anomalous dimension in maximally supersymmetric Yang-Mills theory. Phys. Rev. **D75**, 085010 (2007, hep-th/0610248)
23. J. Drummond, J. Henn, G. Korchemsky, E. Sokatchev, Dual superconformal symmetry of scattering amplitudes in N=4 super-Yang-Mills theory. Nucl. Phys. **B828**, 317–374 (2010, arXiv:0807.1095)

24. N. Berkovits, J. Maldacena, Fermionic T-Duality, dual superconformal symmetry, and the amplitude/Wilson loop connection. JHEP **0809**, 062 (2008, arXiv:0807.3196)
25. N. Beisert, T-Duality, dual conformal symmetry and integrability for strings on AdS(5) x S**5. Fortsch. Phys. **57**, 329–337 (2009, arXiv:0903.0609)
26. J.M. Drummond, J.M. Henn, J. Plefka, Yangian symmetry of scattering amplitudes in N=4 super Yang-Mills theory. JHEP **0905**, 046 (2009, arXiv:0902.2987)
27. J. Drummond, L. Ferro, Yangians Grassmannians and T-duality. JHEP **1007**, 027 (2010, arXiv:1001.3348)
28. L. Dolan, C.R. Nappi, E. Witten, A relation between approaches to integrability in superconformal Yang-Mills theory. JHEP **0310**, 017 (2003, hep-th/0308089)
29. L. Dolan, C.R. Nappi, E. Witten, Yangian symmetry in D = 4 superconformal Yang-Mills theory, hep-th/0401243
30. N. Arkani-Hamed, F. Cachazo, C. Cheung, J. Kaplan, A duality for the S matrix. JHEP **1003**, 020 (2010, arXiv:0907.5418)
31. N. Arkani-Hamed, F. Cachazo, C. Cheung, The Grassmannian origin of dual superconformal invariance. JHEP **1003**, 036. (2010, arXiv:0909.0483)
32. L. Mason, D. Skinner, Dual superconformal invariance, momentum twistors and Grassmannians. JHEP **0911**, 045 (2009, arXiv:0909.0250)
33. N. Arkani-Hamed, J.L. Bourjaily, F. Cachazo, S. Caron-Huot, J. Trnka, The all-loop integrand for scattering amplitudes in planar N=4 SYM. JHEP **1101**, 041 (2011, arXiv:1008.2958)
34. J.M. Drummond, J.M. Henn, Simple loop integrals and amplitudes in N=4 SYM. JHEP **1105**, 105 (2011, arXiv:1008.2965)
35. J. Drummond, L. Ferro, The Yangian origin of the Grassmannian integral. JHEP **1012**, 010 (2010, arXiv:1002.4622)
36. M. Bullimore, D. Skinner, Descent equations for superamplitudes, arXiv:1112.1056
37. S. Caron-Huot, S. He, Jumpstarting the all-loop S-Matrix of planar N=4 super Yang-Mills. JHEP **1207**, 174 (2012, arXiv:1112.1060)
38. J. Drummond, J. Henn, G. Korchemsky, E. Sokatchev, Generalized unitarity for N=4 superamplitudes, arXiv:0808.0491
39. J. Drummond, J. Henn, All tree-level amplitudes in N=4 SYM. JHEP **0904**, 018 (2009, arXiv:0808.2475)
40. L. Mason, D. Skinner, Scattering amplitudes and BCFW recursion in twistor space. JHEP **1001**, 064 (2010, arXiv:0903.2083)
41. L. Mason, D. Skinner, The complete planar S-matrix of N=4 SYM as a Wilson loop in twistor space. JHEP **1012**, 018 (2010, arXiv:1009.2225)
42. L.F. Alday, J.M. Henn, J. Plefka, T. Schuster, Scattering into the fifth dimension of N=4 super Yang-Mills. JHEP **1001**, 077 (2010, arXiv:0908.0684)
43. G. Georgiou, V.V. Khoze, Tree amplitudes in gauge theory as scalar MHV diagrams. JHEP **0405**, 070 (2004, hep-th/0404072)
44. G. Georgiou, E. Glover, V.V. Khoze, Non-MHV tree amplitudes in gauge theory. JHEP **0407**, 048 (2004, hep-th/0407027)
45. R. Boels, C. Schwinn, CSW rules for a massive scalar. Phys. Lett. **B662**, 80–86 (2008, arXiv:0712.3409)
46. R. Boels, C. Schwinn, CSW rules for massive matter legs and glue loops. Nucl. Phys. Proc. Suppl. **183**, 137–142 (2008, arXiv:0805.4577)
47. A. Brandhuber, B.J. Spence, G. Travaglini, One-loop gauge theory amplitudes in N=4 super Yang-Mills from MHV vertices. Nucl. Phys. **B706**, 150–180 (2005, hep-th/0407214)
48. J. Bedford, A. Brandhuber, B.J. Spence, G. Travaglini, A twistor approach to one-loop amplitudes in N=1 supersymmetric Yang-Mills theory. Nucl. Phys. **B706**, 100–126 (2005, hep-th/0410280)
49. C. Quigley, M. Rozali, One-loop MHV amplitudes in supersymmetric gauge theories. JHEP **0501**, 053 (2005, hep-th/0410278)
50. P. Mansfield, The lagrangian origin of MHV rules. JHEP **0603**, 037 (2006, hep-th/0511264)

51. J.H. Ettle, T.R. Morris, Structure of the MHV-rules Lagrangian. JHEP **0608**, 003 (2006, hep-th/0605121). Makes major improvements on withdrawn paper hep-th/0605024
52. R. Boels, L. Mason, D. Skinner, From twistor actions to MHV diagrams. Phys. Lett. **B648**, 90–96 (2007, hep-th/0702035)
53. M. Bullimore, L. Mason, D. Skinner, MHV diagrams in momentum twistor space. JHEP **1012**, 032 (2010, arXiv:1009.1854)
54. M. Bullimore, MHV diagrams from an all-line recursion relation, arXiv:1010.5921
55. A. Brandhuber, G. Travaglini, Quantum MHV diagrams, hep-th/0609011

Chapter 4
On-Shell Recursion

Amongst the most efficient methods for generating tree-level scattering amplitudes in gauge theories are on-shell recursion relations, introduced in the remarkable papers [1, 2]. These first on-shell recursion relations involved a particular analytic continuation of the external momenta and are now known as the BCFW recursion relations. However, the on-shell recursion method is very general and many such recursion relations have since been constructed. It essentially depends only on the analytic behaviour scattering amplitudes as a function of the external momenta and the power of complex analysis. Amplitudes are constructed recursively while remaining completely on-shell.

The on-shell recursion relations of [1, 2] were originally introduced for the tree-level scattering of gluons, and now have a broad range of applications including theories containing massive particles [3, 4] and to rational one–loop amplitudes [5, 6]. They have been extended to compute tree-level superamplitudes [7, 8] and loop integrands in planar $\mathcal{N} = 4$ supersymmetric Yang-Mills theory [9, 10]. On-shell recursion relations can also be formulated in momentum twistor space, and each step in the recursion relation has been shown to be invariant under the Yangian symmetry, providing one way to prove the invariance of all tree-level superamplitudes and loop integrands [9].

In this chapter we consider another recursion relation called the 'all-line' recursion relation, which is intimately related to the MHV diagram formalism and was introduced for tree-level superamplitudes in [11, 12]. We reexpress this recursion relation in momentum twistor space and extend it to all loop integrands. Also, we show that the complete solution of this recursion relation is precisely the MHV diagram expansion, thus proving that it indeed computes all four-dimensional loop integrands. This chapter is based on reference [13].

M. R. Bullimore, *Scattering Amplitudes and Wilson Loops in Twistor Space*, Springer Theses, DOI: 10.1007/978-3-319-00909-4_4,

4.1 General Considerations

In order to derive an on-shell recursion relation, we choose any complex deformation of the incoming supermomenta

$$\lambda_j \to \lambda_j(z) \qquad \tilde{\lambda}_j \to \tilde{\lambda}_j(z) \qquad \eta_j \to \eta_j(z)\,, \tag{4.1}$$

which preserves the on-shell and momentum conservation conditions. The tree-level superamplitudes and loop integrands then become meromorphic functions of the deformation parameter z. We then consider the contour integral

$$\oint_C \frac{dz}{z}\mathcal{A}(z) = 0 \tag{4.2}$$

whose contour C surrounds the point at infinity. There one simple pole at the origin whose residue is the original superamplitude or integrand. Then there are additional simple poles when

1. Sums of external momenta go on-shell: $(p_i + \cdots + p_{j-1})^2 = (x_i - x_j)^2 \to 0$.
2. Internal momentum channels go on-shell $(x - x_j)^2 \to 0$.

On these simple poles, superamplitudes factorize according to standard LSZ reduction formulae. This analytic behaviour is essentially the only input for deriving recursion relations. Provided it can be shown independently that there is no pole at infinity $z \to \infty$, we obtain a recursion relation of the schematic form

$$\mathcal{A} = \sum_{L,R} \int \mathrm{d}^4\eta_I \, \mathcal{A}_L(I) \, \frac{1}{P_I^2} \, \mathcal{A}_R(-I) + \sum_I \frac{1}{P_I^2} \int \mathrm{d}^4\eta_I \, \mathcal{A}(I,-I)\,. \tag{4.3}$$

The right-hand side receives contributions from 1. 'factorization terms' which appear for both tree-level superamplitudes and loop integrands and 2. 'forward terms' which appear only for loop integrands. This is illustrated in Fig. (4.1).

Since momentum twistors manifest momentum conservation and the on-shell conditions, they are completely unconstrained variables. Thus for constructing recursion relations we may choose *any* deformation of the momentum twistor polygon

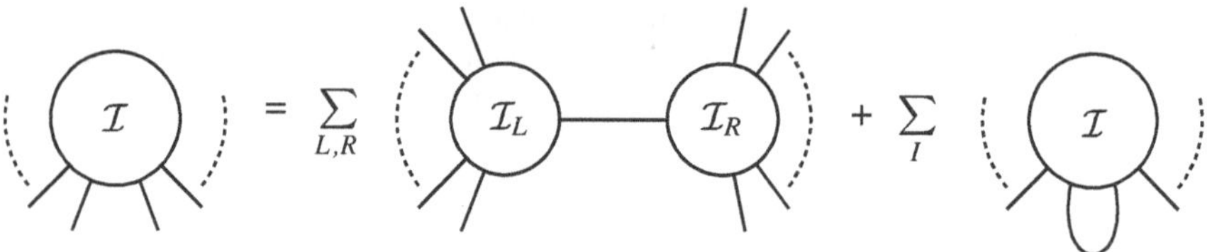

Fig. 4.1 A graphical representation of recursion relations for loop integrands

$$\mathcal{Z}_j \longrightarrow \mathcal{Z}_j(z) \tag{4.4}$$

There are two particularly convenient choices of momentum twistor deformations that have been employed in the literature:

1. All-line deformation: $\mathcal{Z}_j(z) = \mathcal{Z}_j - zr_j\mathcal{Z}_*$.
2. BCFW deformation: $\mathcal{Z}_n(z) = \mathcal{Z}_n - z\mathcal{Z}_{n-1}$.

The BCFW recursion relation leads to fewer terms and manifest Yangian symmetry of tree-level superamplitudes and loop integrands. On the other hand, the all-line recursion relation manifests cyclic and dual superconformal symmetry and leads directly to the MHV diagram formalism.

4.2 All-Line Recursion

4.2.1 All-Line Deformation

The all-line deformation is

$$\mathcal{Z}_j(z) = \mathcal{Z}_j + z\, r_j \mathcal{Z}_* \tag{4.5}$$

where the $r_j \in \mathbb{C}^*$ and $\mathcal{Z}_*$ is an arbitrary reference twistor. In momentum space it is convenient to choose components $\mathcal{Z}_* = (0, \zeta^{\dot{\alpha}}, 0)$ so that the deformation affects only the secondary components

$$\mu_j^{\dot{\alpha}} \to \mu_j^{\dot{\alpha}} + z\, r_j \zeta^{\dot{\alpha}}\,. \tag{4.6}$$

The region momenta are then shifted by

$$x_i^{\alpha\dot{\alpha}} \to x_i^{\alpha\dot{\alpha}} + z\, q_i^{\alpha} \zeta^{\dot{\alpha}} \tag{4.7}$$

where

$$q_i^{\alpha} \equiv \frac{r_i \lambda_{i-1}^{\alpha} - r_{i-1}\lambda_i^{\alpha}}{\langle i-1\, i\rangle} \tag{4.8}$$

It will be important to the choose the r_j such that all region momenta are shifted in order to capture all possible factorization channels in the recursion relation.

4.2.2 Behaviour at Infinity

In order to obtain a recursion relation, one must ensure there are no additional poles as $z \to \infty$ since such terms would not correspond to any factorization channel and cannot be determined a priori. In this limit, all momentum twistors are sent towards the reference twistor Z_* in a multiple collinear limit. Since the reference twistor is an arbitrary choice, the recursion relation should not receive any contributions from such a configuration. Indeed, we will now show that dual superconformal symmetry ensures that integrands of ℓ-loop N^{k}MHV amplitudes scale like $\mathcal{O}(z^{-k-\ell-1})$ and that tree-level superamplitudes decay at least as rapidly as $\mathcal{O}(z^{-k})$ in the limit that $z \to \infty$.

Tree-level superamplitudes in planar $\mathcal{N} = 4$ SYM can be expressed as sums of products of dual superconformal brackets $[\,,\,,\,,\,,\,]$ involving only the external momentum twistors [14]. Since the dual conformal four bracket $\langle\,,\,,\,,\,\rangle$ is completely antisymmetric in its arguments,

$$\langle\,,\,,\,,\,\rangle \sim \mathcal{O}(z)\,. \tag{4.9}$$

for any choice of momentum twistor arguments. The fermionic delta-functions of the form $\delta^{0|4}(\ldots)$ then scale uniformly as $\mathcal{O}(z^4)$ and thus for any dual superconformal bracket we have

$$[\,,\,,\,,\,,\,] \sim \mathcal{O}(z^{-1})\,. \tag{4.10}$$

For N^{k}MHV superamplitudes, the answer is expressed as a sum of products of k dual superconformal brackets $[\,,\,,\,,\,,\,]$ with potentially shifted arguments. Hence the tree-level superamplitudes behave like $O(z^{-k})$ in agreement with the scaling behaviour expected on general grounds [12, 15].

The supersymmetric loop integrals in the definition of the loop integrand contain the fermionic integrals

$$\int \prod_j \mathrm{d}^4\chi_{A_j}\mathrm{d}^4\chi_{B_j} \tag{4.11}$$

over the fermionic components of the internal momentum twistors $\mathcal{Z}_{A_j}$ and $\mathcal{Z}_{B_j}$. Once these fermionic integrals have been performed, the remaining bosonic loop integrands of N^{k}MHV superamplitudes have grassmann degree $4k$. The bosonic loop integrand can then be expanded in a basis of local integrands with unit leading singularity. The coefficients in this expansion are products of k dual superconformal brackets $[\,,\,,\,,\,,\,]$ and behave as $\mathcal{O}(z^{-k})$ under the all-line deformation.

Now consider the behaviour of local integrands with unit leading singularity.[1] Dual conformal invariance requires that they are constructed from momentum twistor four-brackets $\langle\,,\,,\,,\,\rangle$ with an overall weight zero in all external momentum twistors $\mathcal{Z}_i$ and weight -4 in the loop momentum twistors $\mathcal{Z}_A$ and $\mathcal{Z}_B$.

Any such one-loop integrand necessarily has the form [9]

$$\frac{\langle a, b, Y_1\rangle \ldots \langle a, b, Y_{n-4}\rangle}{\langle a, b, 1, 2\rangle\langle a, b, 2, 3\rangle \ldots\ldots \langle a, b, n, 1\rangle}\,, \tag{4.12}$$

[1] The large-z behaviour of loop integrands under the BCFW deformation was considered in detail in reference [10].

where all physical propagators are included in the denominator and $\{Y_1, \ldots, Y_{n-4}\}$ are skew twistors. The skew twistors need not be simple and hence do not necessarily correspond to lines in twistor space. The antisymmetric twistors $\{Y_1, \ldots, Y_{n-4}\}$ must together carry weight $2n$ in each external momentum twistor. Thus, expanding each one in a basis of simple bitwistors of the form $Z_i^{[I} Z_j^{J]}$, the overall coefficients contain two dual conformal brackets $\langle\, , \, , \, , \,\rangle$. For example, one term that can appear is the chiral pentagon integrand illustrated in Fig. (4.2),

$$\frac{\langle a, b, 1, 4\rangle\langle 5, 1, 2, 3\rangle\langle 2, 3, 4, 5\rangle}{\langle a, b, 1, 2\rangle\langle a, b, 2, 3\rangle\langle a, b, 3, 4\rangle\langle a, b, 4, 5\rangle\langle a, b, 5, 1\rangle}. \tag{4.13}$$

Four-brackets corresponding to internal propagators behave as

$$\langle a, b, \, , \,\rangle \sim \mathcal{O}(z) \tag{4.14}$$

and therefore by counting weights, the one-loop integrands of unit leading singularity behave as $\mathcal{O}(z^{-2})$. Now combining this with the behaviour of the coefficients, each term in the expansion has behavior $\mathcal{O}(z^{-k-2})$. Thus the complete one-loop integrand behaves by at least the same power.

Since both dual conformal brackets of the form $\langle\, , \, , \, , \,\rangle$ and $\langle a\, b, \, , \,\rangle$ behave as $\mathcal{O}(z)$ then increasing the number of loops can only improve the $z \to \infty$ behavior of loop integrand. An example is the two-loop penta-box integrand in see Fig. (4.2),

$$\frac{\langle 3, 4, 5, 1\rangle\langle 4, 5, 1, 3\rangle\langle a, b|(5, 1, 2) \cap (2, 3, 4)\rangle}{\langle a, b, 5, 1\rangle\langle a, b, 1, 2\rangle\langle a, b, 2, 3\rangle\langle a, b, 3, 4\rangle\langle a, b, c, d\rangle\langle c, d, 3, 4\rangle\langle c, d, 4, 5\rangle\langle c, d, 5, 1\rangle}. \tag{4.15}$$

Expanding the numerator

$$\langle ab|(5, 1, 2) \cap (2, 3, 4)\rangle = \langle a, 5, 1, 2\rangle\langle b, 2, 3, 4\rangle - \langle b, 5, 1, 2\rangle\langle a, 2, 3, 4\rangle \tag{4.16}$$

it is clear that the integrand falls away as $\mathcal{O}(z^{-3})$. An extension of the above argument shows all local integrands of unit leading singularity fall away as $\mathcal{O}(z^{-\ell-1})$.

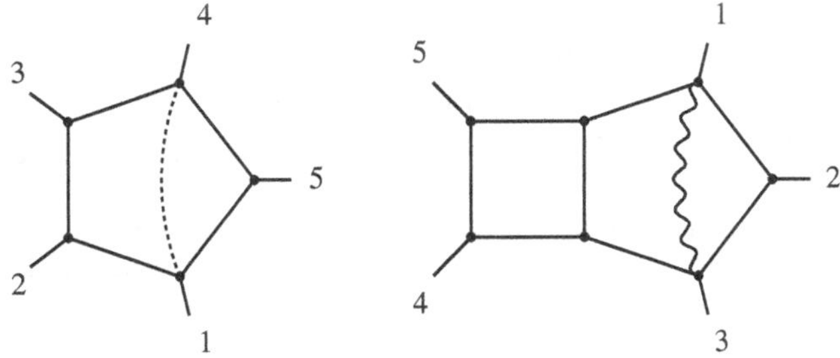

Fig. 4.2 Examples of local integrands with unit leading singularities

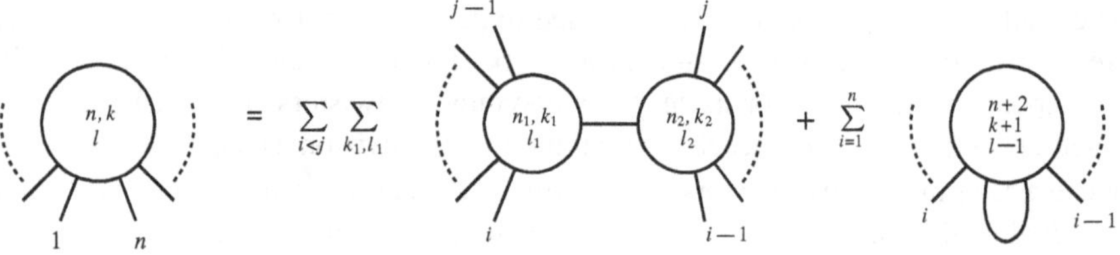

Fig. 4.3 A graphical representation of the all-line recursion relation for loop integrands

Combining this scaling with that of the coefficients, each term in a local expansion falls away as $\mathcal{O}(z^{-k-\ell-1})$ and hence the full integrand behaves at least this well.

4.2.3 The Recursion Relation

The full recursion relations for the momentum space loop integrand are

$$\begin{aligned}\mathcal{A}_{n,k}^{(\ell)} = & \sum_{i,j,k_1,\ell_1} \int \mathrm{d}^4\eta_I \, \mathcal{A}_{n_1,k_1}^{(\ell_1)}(i,\ldots,j-1,I;z_I) \frac{1}{(x_i-x_j)^2} \mathcal{A}_{n_2,k_2}^{(\ell_2)}(j,\ldots,i-1,-I;z_I) \\ & + \sum_{i=1}^{n} \frac{1}{(x-x_i)^2} \int \mathrm{d}^4\eta_I \, \mathcal{A}_{n+2,k+1}^{(\ell-1)}(i,\ldots,i-1,I,-I;z_I)\,. \end{aligned} \tag{4.17}$$

The channels going on shell are labelled with subscripts I and the notation $\mathcal{A}(\ldots;z_I)$ means that all external region momenta are deformed and evaluated on the relevant pole z_I. The external legs denoted by I and $-I$ correspond to the variables

$$I = \{\lambda_I, \eta_I\} \qquad -I = \{-\lambda_I, \eta_I\} \tag{4.18}$$

where λ_I is a left-handed spinor associated with the off-shell momentum P_I flowing in the channel I. This will be discussed further below where we deal separately with the two contributions in Eq. (4.17).

The summation ranges in the first line of Eq. (4.17) are as follows. There is a sum over the numbers of external legs $1 \leq i < j \leq n$ where $n_1 + n_2 = n + 2$. In addition there is a summation over the allowed grassmann degrees $0 \leq k_1 \leq k-1$ where $k_2 + k_2 = k + 1$, and number of loops $0 \leq \ell_1 \leq \ell$ where $\ell_1 + \ell_2 = \ell$. Although the internal loop regions have not been denoted explicitly, the whole expression must be symmetrized over the assignment of region momenta to internal regions (Fig. 4.3).

For tree-level superamplitudes, the only poles are from propagators $1/(x_i - x_j)^2$, and Eq. (4.17) reduces to the simpler all-line recursion relation

$$\mathcal{A}_{n,k}^{(0)} = \sum_{i,j,k_1} \int \mathrm{d}^4\eta_I \, \mathcal{A}_{n_1,k_1}^{(0)}(i,\ldots,j-1,I;z_I) \, \frac{1}{(x_i-x_j)^2} \, \mathcal{A}_{n_2,k_2}^{(0)}(j,\ldots,i-1,-I;z_I) \,. \tag{4.19}$$

We will also write the all-line recursion relation in momentum twistor space using the techniques developed in [9]. This has the advantage of generating MHV diagrams in the dual superconformally invariant manner. The full recursion relations for the momentum twistor integrand are

$$\begin{aligned}\mathcal{A}_{n,k}^{(\ell)} = &\sum_{i,j,k_1,\ell_1} [*,i-1,i,j-1,j] \, \mathcal{A}_{n_1,k_1}^{(\ell_1)}(Z_I,i,\ldots,j-1;z_I) \, \mathcal{A}_{n_2,k_2}^{(\ell_2)}(Z_I,j,\ldots,i-1;z_I) \\ &+ \sum_{i=1}^{n} \oint_{GL(2)} [*,i-1,i,a,b] \, \mathcal{A}_{n+2,k+1}^{(\ell-1)}(i,\ldots,i-1,a,b';z_I) \,. \end{aligned} \tag{4.20}$$

where the summations are the same as the preceding paragraph. The momentum twistor $\mathcal{Z}_b'$ is defined by the intersection point $\langle *,i-1,i,[\mathcal{Z}_a\rangle\,\mathcal{Z}_b]$ and the momentum twistor $\mathcal{Z}_I$ is a natural point in the geometry described below. The external momentum twistors $\mathcal{Z}_i$ are all shifted and evaluated on the relevant pole z_I, and for multi-loop amplitudes, the expression must be symmetrized over the possible lines $(ab)_m$.

Factorization Terms

Firstly, there are poles from propagators bounding two external regions, which arise for both tree-level superamplitudes and loop integrands. Consider the propagator $1/P_I^2 = 1/(x_i-x_j)^2$ with momentum in the channel $I = \{i,\ldots,j-1\}$. The pole occurs when the shifted momentum $P_I(z) = x_{ij}(z)$ becomes null:

$$z_I = -\frac{x_{ij}^2}{2\langle q_I|x_{ij}|\zeta]} \tag{4.21}$$

where $q_I^\alpha = q_i^\alpha - q_j^\alpha$. Standard arguments ensure factorization into integrands containing fewer numbers of legs and numbers of loops—see Fig. 4.4.

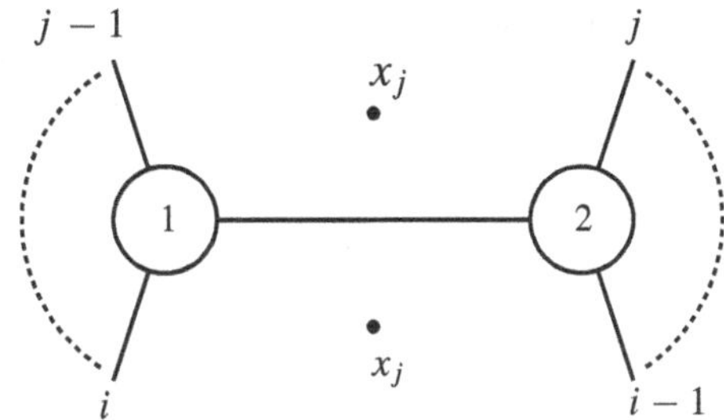

Fig. 4.4 Factorization of an integrand along the channel $(x_i-x_j)^2 \to 0$

In momentum space the contribution to the recursion relation is

$$\int d^4\eta_I \, A^{(\ell_1)}_{n_1,k_1}(i,\ldots,j-1,I;z_I)\,\frac{1}{(x_i-x_j)^2}\,A^{(\ell_2)}_{n_2,k_2}(j,\ldots,i-1,-I;z_I) \quad (4.22)$$

where all external region momenta are shifted and evaluated on the pole z_I. The on-shell momentum $P_I(z_I)$ may be written $\lambda_I\tilde{\lambda}_I$ and contracting with the reference spinor $\zeta^{\dot{\alpha}}$ we find that the holomorphic spinor associated with the off-shell propagator is $\lambda_I = P_I\,|\zeta]/[\tilde{\lambda}_I\,\zeta]$. Since the product of integrands together with the fermion measure $d^4\eta_I$ are invariant under little group transformations $(\lambda_I,\eta_I)\to(t\lambda_I,t^{-1}\eta_I)$, we may rescale and define

$$\lambda_I = P_I|\zeta] = (x_i-x_j)|\zeta] \quad (4.23)$$

This is the CSW prescription for the spinor assigned to the off-shell momentum P_I [16]. The notation I and $-I$ in Eq. (4.22) is shorthand for the variables $\{\lambda_I,\eta_I\}$ and $\{-\lambda_I,\eta_I\}$.

We now consider the momentum twistor geometry associated with this pole. First consider that there are no shifts of the momentum twistors. The two regions x_i and x_j correspond to lines X_i and X_j in momentum twistor space which are generically skew. However, the reference twistor determines a unique line intersecting X_i and X_j which then defines two distinguished points on those lines (see Fig. 4.5)

$$\mathcal{Z}_{ij} = \langle * \, i-1, i, [\,j-1\rangle\,\mathcal{Z}_j] \qquad \mathcal{Z}_{ji} = \langle * \, j-1, j, [\,i-1\rangle\,\mathcal{Z}_i] \quad (4.24)$$

which are the intersection points $(i-1,i)\cap(j-1,j\,*)$ and $(j-1,j)\cap(i-1,i\,*)$ in momentum twistor space. The supersymmetric extension of this statement is valid on the support of the fermionic delta-functions in the superconformal bracket $[\,*\,,i-1,i,j-1,j]$ associated with $1/(x_i-x_j)^2$.

There is one additional distinguished point in the geometry since three points on a line automatically define a fourth $\mathcal{Z}_I$ which is equi-anharmonic. This means that the four points have cross-ratio -1. The components of $\mathcal{Z}_I$ are

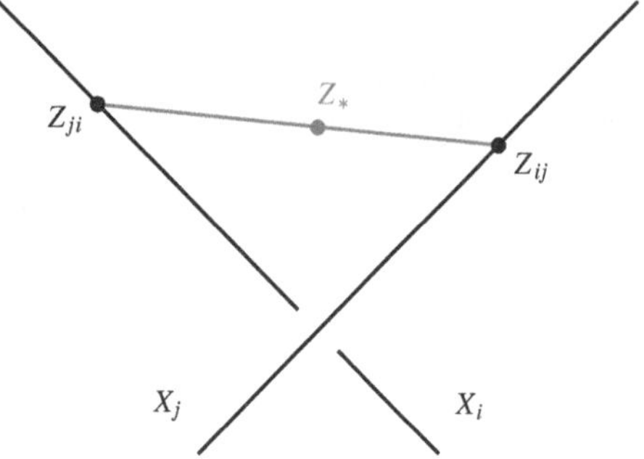

Fig. 4.5 The geometry of poles in the loop integrand in momentum twistor space

$$\begin{aligned}\mathcal{Z}_I &= \mathcal{Z}_{ij} + \frac{1}{2}\langle i-1, i, j-1, j\rangle \mathcal{Z}_* \\ &= \mathcal{Z}_{ji} - \frac{1}{2}\langle i-1, i, j-1, j\rangle \mathcal{Z}_* \,.\end{aligned} \tag{4.25}$$

The equality of the bosonic components follows from the linear dependence of five bosonic momentum twistors and equality of the fermionic components follows on the support of the dual superconformal bracket $[\,*\,, i-1, i, j-1, j\,]$. This special point in the geometry plays an important role in the recursion relation.

Now consider the all-line shift of momentum twistors:

$$\mathcal{Z}_j(z) = Z_j + z\, r_j\, Z_* \tag{4.26}$$

under which the line $X_j(z)$ sweeps out the plane $(j-1, j, *)$. There are poles in the loop integrand when non-adjacent lines $X_i(z)$ and $X_j(z)$ intersect and it is straightforward to show (for example, by using incidence relations) that the intersection is the equi-anharmonic point $\mathcal{Z}_I$. The primary component of $\mathcal{Z}_I$ is then the holomorphic CSW spinor λ_I associated with the off-shell propagator.

The contribution to the recursion relation from this pole is then

$$[\,*\,, i-1, i, j-1, j]\, \mathcal{A}^{(\ell_1)}_{n_1,k_1}(I, i, \ldots, j-1; z_I)\, \mathcal{A}^{(\ell_2)}_{n_2,k_2}(I, j, \ldots, i-1; z_I) \tag{4.27}$$

where the external momentum twistors are evaluated on the parameter z_I where the pole occurs. Note that the propagator contributes a dual superconformal bracket $[\,*\,, i{-}1, i, j{-}1, j]$. This is a direct consequence of translating a factorization channel into momentum twistor space [17].

Forward Terms

There are also simple poles from propagators bounding an internal and an external region. Consider the propagator $1/(x-x_i)^2$ where x is an internal region. This occurs when the four-momentum $P_I(z) = (x - x_i(z))$ becomes null at

$$z_I = \frac{(x-x_i)^2}{2\langle q_i|(x-x_i)|\zeta]} \tag{4.28}$$

Standard arguments ensure factorization and the residue corresponds a forward limit in the channel $(x - x_i)$ of an integrand with one less loop (see Fig. 4.6).

The contribution to recursion relations from this pole is

$$\int \mathrm{d}^4\eta_I\, \frac{1}{(x-x_i)^2}\, \mathcal{A}^{(\ell-1)}_{n+2,k+1}(i, \ldots, i-1, I, -I; z_I) \tag{4.29}$$

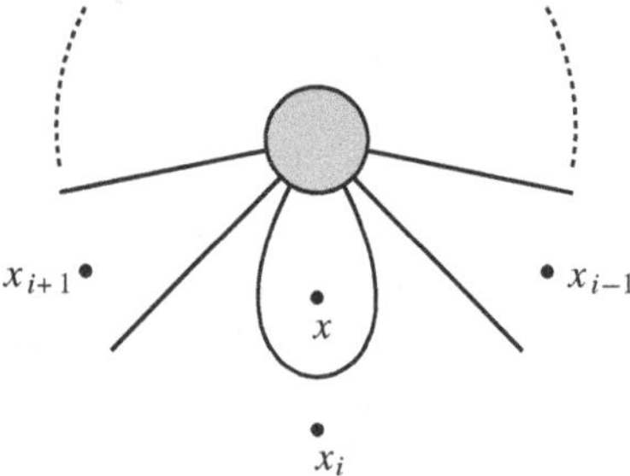

Fig. 4.6 The geometry of the forward limit terms in the recursion relation.

where the external region momenta are all shifted and evaluated on the pole z_I and the notation I and $-I$ is shorthand for $\{\lambda_I, \eta_I\}$ and $\{-\lambda_I, \eta_I\}$.

This is an equation for the loop integrand, and consequently the position of the pole $z_I = z_I(x)$ is a function of the loop momentum. Again, the on-shell momentum $P_I(z_I)$ in the propagator may be written as $\lambda_I \tilde{\lambda}_I$ and contracting with the reference spinor we find the holomorphic spinor $|\lambda_I\rangle = P_I\,|\zeta]/[\tilde{\lambda}_I \zeta]$. However, Eq. (4.29) is invariant under little group transformations $(\lambda_I, \eta_I) \to (t\lambda_I, t^{-1}\eta_I)$ and we may equally use the spinor

$$|\lambda_I\rangle = P_I|\zeta] = (x - x_i)|\zeta]\,. \tag{4.30}$$

This is again the CSW prescription for assigning a holomorphic spinor to the off-shell momentum $P_I(z)$ but since we are dealing with the loop integrand the spinor $\lambda_I = \lambda_I(x)$ depends on the loop momentum.

When $(x - x_i)^2 \to 0$ the same configuration is obtained as the forward limit of a null polygon with two additional cusps $\{x', x\}$ by sending x' towards the original cusp x_i while x becomes a new internal region. In momentum twistor space, a new line $X = (a, b)$ intersects the component X_i and additional momentum twistors $\mathcal{Z}_a$ and $\mathcal{Z}_b$ tend towards the intersection $X \cap X_i$. The situation is illustrated in Fig. 4.7.

Since the deformed line $X_i(z)$ sweeps out the plane $(i-1, i, *)$ this happens at the intersection point $b' \equiv (a, b) \cap (i-1, i, *)$ with components

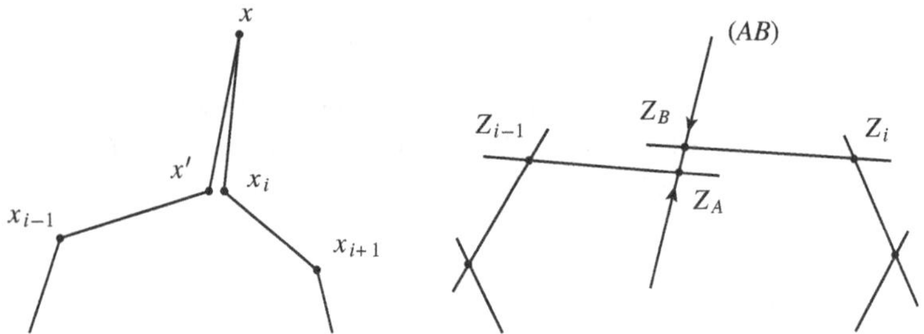

Fig. 4.7 The region space and momentum twistor geometry of a forward limit

$$\mathcal{Z}'_b = \langle *, i-1, i, [\mathcal{Z}_a \rangle\, \mathcal{Z}_b] \,. \tag{4.31}$$

The primary component of $Z_{b'}$ is then the holomorphic spinor $(x - x_i)|\zeta]$ associated with the propagator $1/(x - x_i)^2$. The forward limit is again implemented by sending the additional loop momentum twistors $\mathcal{Z}_a$ and $\mathcal{Z}_b$ along the line X towards the intersection point $\mathcal{Z}'_b$.

The forward limit may is implemented in momentum twistor space following the technique introduced in [9]. The loop integral $\mathrm{d}^{4|8}X$ over chiral superspace is simply an integration over lines in $\mathbb{CP}^{3|4}$. Since each line is determined by a pair of momentum twistors $\mathcal{Z}_a$ and $\mathcal{Z}_b$, the same integration is performed by integrating twice over $\mathbb{CP}^{3|4}$, and dividing by the volume of GL(2) that translates $\mathcal{Z}_a$ and $\mathcal{Z}_b$ along the line X. To make this concrete, we can write

$$\mu^{\dot\alpha}_{(a)} = i x^{\alpha\dot\alpha} \lambda_{(a)\alpha} \qquad\qquad \mu^{\dot\alpha}_{(b)} = i x^{\alpha\dot\alpha} \lambda_{(b)\alpha} \tag{4.32}$$

(with a similar equation for the fermionic components) so that the holomorphic spinors λ_a and λ_b become homogeneous coordinates on $\mathbb{CP}^1 \times \mathbb{CP}^1$ parametrizing the positions of $\mathcal{Z}_a$ and $\mathcal{Z}_b$ on the line X. The integration measure then decomposes as follows:

$$\mathrm{D}^{3|4} Z_a \, \mathrm{D}^{3|4} Z_b \;=\; \mathrm{d}^{4|8} X \; \langle \lambda_a \lambda_b \rangle^2 \langle \lambda_a \mathrm{d}\lambda_b \rangle \langle \lambda_b \mathrm{d}\lambda_b \rangle \,. \tag{4.33}$$

When the integrand depends on $\mathcal{Z}_a$ and $\mathcal{Z}_b$ only through the line $X = (a, b)$ then the only poles in the spinors λ_a or λ_b appears when $\lambda_a = \lambda_b$. Hence we choose the contour $\lambda_b = \overline{\lambda}_a$ which identifies antipodal points in $\mathbb{CP}^1 \times \mathbb{CP}^1$ and integration is simply performed and yields $(2\pi i)^2$ leaving the required measure $\mathrm{d}^{4|8}X$. In this case we can write

$$\mathrm{d}^{4|8} X = \frac{\mathrm{d}^{4|4} Z_A \mathrm{d}^{4|4} Z_B}{\mathrm{vol\,GL}(2)} \,. \tag{4.34}$$

However, we will consider integrands that have spurious poles in λ_a and λ_b arising from denominator factors such as $\langle a, 1, 2, 3\rangle$ or $\langle b, 2, 3, 4\rangle$. In this case, a different contour must be chosen and this provides and opportunity to perform the forward limit algebraically in momentum twistor space. From the momentum twistor geometry in Fig. 4.7, the contour must be chosen so $\mathcal{Z}_a$ and $\mathcal{Z}_b$ are sent towards the intersection point $\mathcal{Z}'_b$.

The contribution to the recursion relation in momentum twistor space is then

$$\oint_\Gamma \mathrm{d}\mu \; [a, b, i-1, i, \, *] \; \mathcal{A}^{(\ell-1)}_{n+2,k+1}(i \ldots, i-1, a, b'; z_I) \tag{4.35}$$

where the external momentum twistors are deformed and evaluated on the position of the pole z_I. Note that the propagator again contributes a dual superconformal bracket $[a, b, i-1, i, \, *]$. As described in the previous paragraph, the contour Γ is chosen

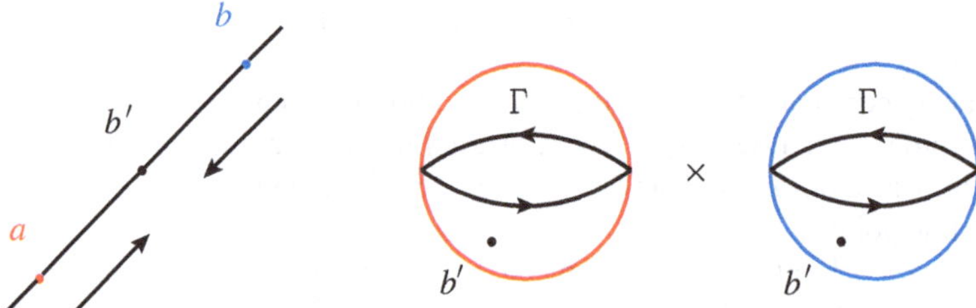

Fig. 4.8 An illustration of the contour Γ that implements the forward limit in momentum twistor space

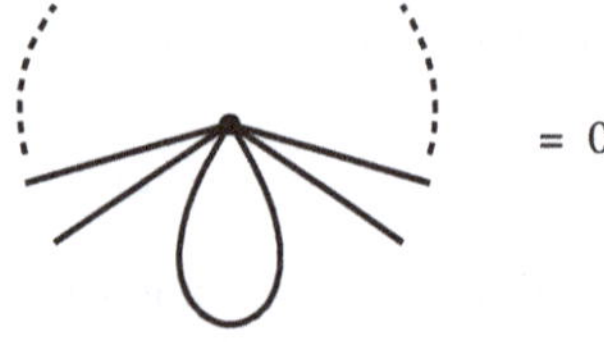

Fig. 4.9 The forward limit of an MHV vertex vanishes

so that $\mathcal{Z}_a$ and $\mathcal{Z}_b$ are sent to the intersection point $\mathcal{Z}'_b$. This contour is illustrated in Fig. 4.8.

Vanishing Forward Terms

It will be important that the forward limit between two adjacent legs on the same MHV vertex vanishes. In such forward limits, there is a fermionic integration $\int d^4\eta_I$ summing over the supermultiplet of states propagating through the channel. While component amplitudes diverge when $\lambda_1 \propto \lambda_2$, the forward limit of a tree-level MHV superamplitude (3.5) vanishes as $\mathcal{O}(\langle\lambda_1\lambda_2\rangle)$ due to a supersymmetric cancellation once the fermionic integration $\int d^4\eta_I$ has been performed Fig. 4.9.

4.3 Some Examples

Here we present some tree-level examples of the momentum twistor recursion relation for tree-level superamplitudes,

$$\mathcal{A}^{(0)}_{n,k}(1,\ldots,n) = \sum_{i,j,k_1} [*, i-1, i, j-1, j]\, \mathcal{A}^{(0)}_{n_1,k_1}(Z_I, i, \ldots, j-1; z_I)$$
$$\mathcal{A}^{(0)}_{n_2,k_2}(Z_I, j, \ldots, i-1; z_I) \tag{4.36}$$

where the summation ranges are $0 \leq i < j \leq n$ and $0 \leq k_1 < k - 1$ with $k_1 + k_2 = k + 1$ and $n_1 + n_2 = n + 2$.

4.3.1 NMHV Tree

For NMHV superamplitudes, the only terms in the recursion relation (4.36) contain two MHV superamplitudes. Summing over all factorization channels we have

$$\mathcal{A}^{(0)}_{n,1}(1,\ldots,n) = \sum_{1\leq i<j\leq n} [*, i-1, i, j-1, j] \tag{4.37}$$

which is immediately the MHV vertex expansion in momentum twistor space.

4.3.2 N²MHV

Terms in the recursion relation (4.36) now contain an MHV and an NMHV superamplitude and the momentum twistor recursion relation becomes (see Fig. 4.10):

$$\mathcal{A}^{(0)}_{n,2}(1,\ldots,n) = \sum_{1\leq i<j\leq n} [*, i-1, i, j-1, j]\, \mathcal{A}^{(0)}_{n_2,1}(I, j, \ldots, i-1; z_I) \tag{4.38}$$

where the right hand superamplitude has $n_2 \equiv i - j$ particles. The momentum twistors are all shifted and evaluated on the relevant shift parameter z_I.

We now expand the NMHV superamplitude using the solution (4.37). All superamplitudes are now MHV and hence the all-line deformation only effects the propagators in the expansion. Thus each term corresponds to an MHV diagram with one propagator shifted and evaluated on the pole of the other. Considering the diagram with channels $P_I = (x_i - x_j)$ and $P_J = (x_k - x_l)$ there are two terms in the recursion relation corresponding to this diagram as illustrated in Fig. (4.11). Their sum is given by the expression

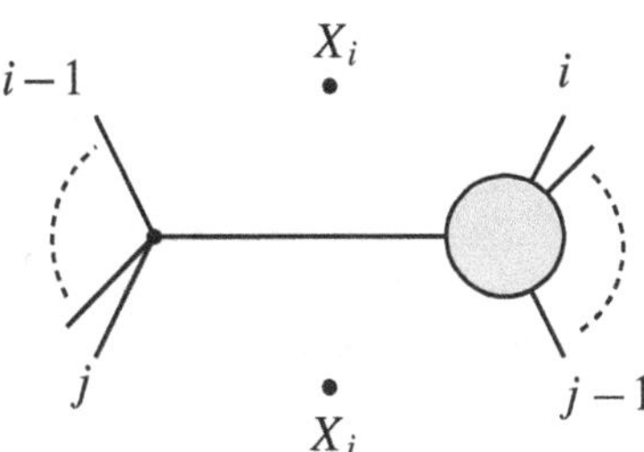

Fig. 4.10 All-line recursion relation for the N^{2}MHV tree-level superamplitude

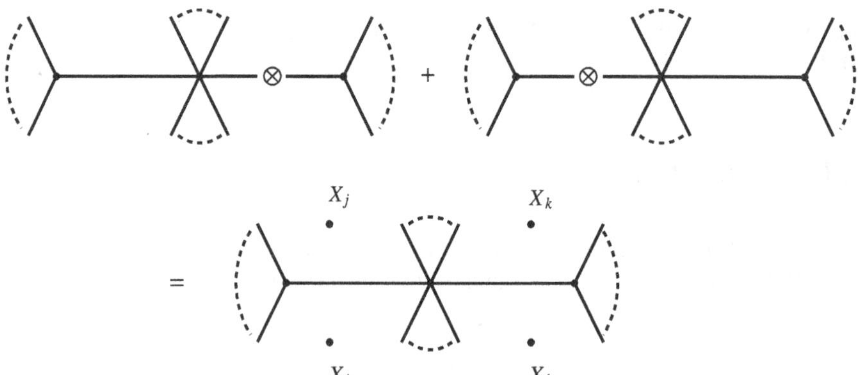

Fig. 4.11 The combination of shifted propagators in the all-line recursion for the N^2 MHV super-amplitude

$$[\,*\,,i-1,i,j-1,j\,]\,[\,*\,,k-1,k,l-1,l\,]\,\Big(\frac{P_I^2}{P_I^2(z_J)}+\frac{P_J^2}{P_J^2(z_I)}\Big). \tag{4.39}$$

Now applying the result of the following contour integral [12]

$$0=\frac{1}{2\pi i}\oint\frac{\mathrm{d}t}{t}\frac{P_I^2P_J^2}{P_I^2(t)P_J^2(t)}=1-\frac{P_I^2}{P_I^2(t_J)}-\frac{P_J^2}{P_J^2(t_I)}, \tag{4.40}$$

the two terms combine to give the correct contribution from an MHV diagram

$$[\,*\,,i-1,i,j-1,j\,]\,[\,*\,,k-1,k,l-1,l\,]. \tag{4.41}$$

This reconstruction of the MHV diagram from the recursion relation is illustrated schematically in Fig. 4.11. Note that we needed to use the same reference twistor at each stage to ensure the cancellation and hence to generate the correct MHV diagram.

There are also boundary diagrams in the channels $P_I = (x_i - x_j)$ and $P_J = (x_k - x_i)$ which are adjacent on the central vertex. The two terms in the recursion relation contributing to this MHV diagram now involve the equi-anharmonic twistors defined in Eq. (4.25). For the channels I and J we have

$$\begin{aligned}\mathcal{Z}_I &= \mathcal{Z}_{ji}+\frac{1}{2}\langle i-1\,i\,j-1\,j\rangle\mathcal{Z}_*\\ \mathcal{Z}_J &= \mathcal{Z}_{ki}+\frac{1}{2}\langle k-1\,k\,i-1\,i\rangle\mathcal{Z}_*\end{aligned} \tag{4.42}$$

where the intersections are

$$\mathcal{Z}_{ji}=(j-1,j)\cap(i-1,i,\,*)\qquad \mathcal{Z}_{ki}=(k-1,k)\cap(i-1,i,\,*). \tag{4.43}$$

Summing both diagrams and again using the residue theorem (4.40) we find the correct expression for the MHV diagram,

$$[\,*\,,i-1,i,\,j-1,\,j\,]\,[\,*\,,k-1,k,i-1,\,Z_{ji}]\,. \tag{4.44}$$

Thus each possible MHV diagram receives contributions from two terms in the recursion relation. Each term is almost the correct MHV diagram expression, except that one propagator is shifted and evaluated on the pole of the other. The two terms then combine in pairs reconstructing the MHV diagram. The same mechanism will occur for all tree-level superamplitudes and loop integrands.

4.3.3 MHV 1-Loop

Before considering the general MHV one-loop integrand, we will understand how the combinatorics of the recursion relation works for the simplest four-point example. For MHV integrands with any numbers of loops, the only terms in the recursion relation are forward terms. For the four-point 1-loop integrand there are four forward terms arising from propagators $(x-x_1)^2$, $(x-x_2)^2$, $(x-x_3)^2$ and $(x-x_4)^2$ as illustrated below.

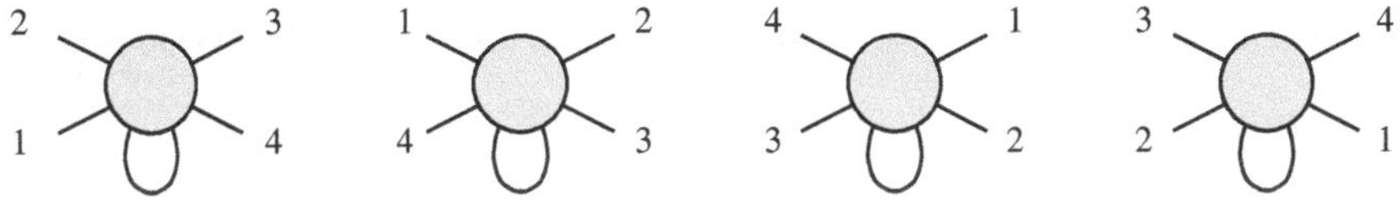

2 3 1 2 4 1 3 4

1 4 4 3 3 2 2 1

These terms are forward limits of six-point NMHV tree-level superamplitudes, each of which is a sum of nine MHV diagrams. However, six involve forward limits of adjacent legs on the same vertex and vanish. The remaining three terms are MHV diagrams of the one-loop MHV integrand except that one propagator has been shifted and evaluated on the pole of the other. For example, for the forward channel $(x-x_1)$ we have

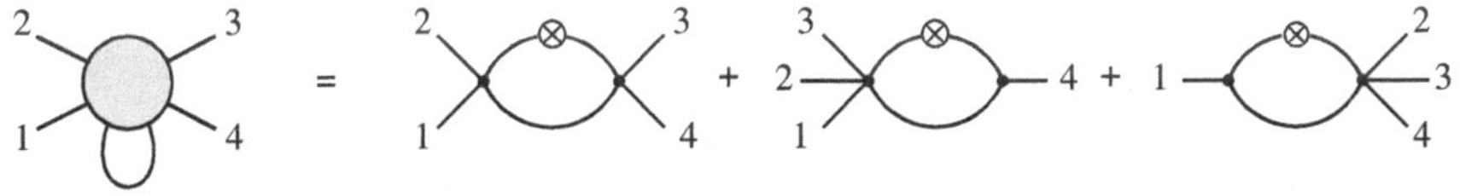

2 3 = 2 3 + 3 2 4 + 1 2 3

1 4 1 4 1 4

There are now $3\times 4 = 12$ terms in total; two terms contribute to each of the six possible MHV diagrams for the four-point one-loop MHV integrand. Choosing the same reference twistor at both stages of the recursion, and using residue theorems of the form

$$\oint \frac{\mathrm{d}z}{z}\frac{1}{(x-x_i(z))^2(x-x_j(z))^2}=0\,, \tag{4.45}$$

the terms combine in pairs to form the six MHV diagrams. For the diagram with channels $(x - x_1)$ and $(x - x_3)$ the combination is illustrated schematically below.

In this way all six MHV diagrams are reconstructed from the recursion relation. We now present the details of this calculation for general one-loop MHV integrands with any number of particles.

Momentum Space

Let us first consider the generic one-loop MHV integrand in momentum space. The only terms in the recursion relation are forward terms and hence the recursion relation becomes (see Fig. 4.12)

$$\mathcal{A}_{n,0}^{(1)} = \sum_{i=1}^{n} \int \mathrm{d}^4\eta_I \frac{1}{(x-x_i)^2} \mathcal{A}_{n+2,1}^{(0)}(1,\ldots,i-1,\{\lambda_I(x),\eta_I\},\{-\lambda_I(x),\eta_I\};z_I(x)) \tag{4.46}$$

where x is the single internal region in the one-loop integrand. The terms in this equation arise from a poles in the integrand where $P_I(t) = (x - x_i(t))$ becomes null

$$z_I(x) = \frac{(x-x_i)^2}{2\langle q_i|(x-x_i)|\zeta]} \,. \tag{4.47}$$

The holomorphic spinor associated with the propagator $1/P_I^2$ is $\lambda_I = P_I|\zeta]$.

We now expand the tree-level NMHV superamplitude using the MHV diagram expansion. Terms involving forward limits of adjacent legs on the same MHV vertex vanish. The remaining terms contain a propagator in an additional channel $(x - x_j)$ and hence the expression for the integrand becomes

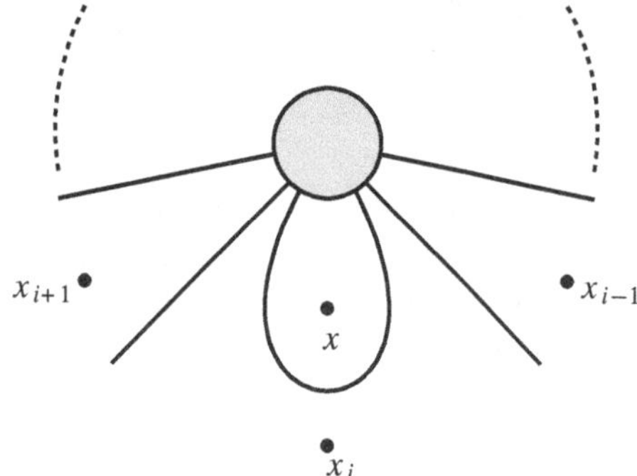

Fig. 4.12 Forward terms contributing to the 1-loop MHV integrand. The shaded circle represents a tree-level NMHV superamplitude

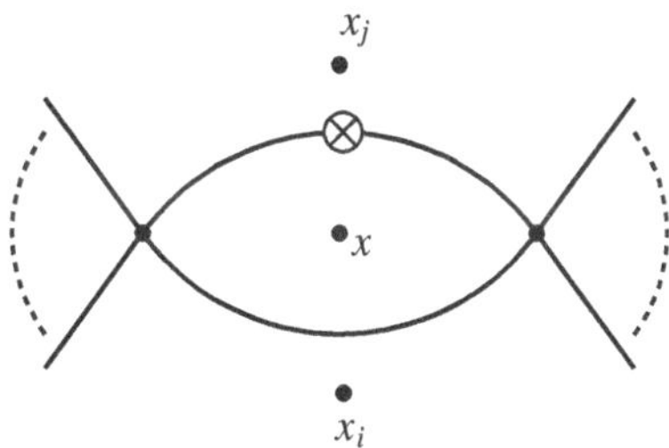

Fig. 4.13 Terms in the all-line recursion relation contributing to the MHV 1-loop integrand

$$\sum_{i=1}^{n}\sum_{j\neq i}\int d^4\eta\, d^4\eta'\, \frac{\mathcal{A}_{\text{MHV}}(\{\lambda,\eta\},i,\ldots,j-1,\{\lambda',\eta'\})\mathcal{A}_{\text{MHV}}(\{\lambda',\eta'\},j\ldots,i-1,\{\lambda,\eta\})}{(x-x_i)^2(x-x_j(t_i))^2} \tag{4.48}$$

where the spinors assigned to each propagator are

$$\lambda^{\alpha}=(x-x_i)^{\alpha\dot{\alpha}}\zeta_{\dot{\alpha}} \qquad \lambda'^{\alpha}=(x-x_j)^{\alpha\dot{\alpha}}\zeta_{\dot{\alpha}}\,. \tag{4.49}$$

Since all external region momenta are shifted and evaluated on the pole z_i, the second propagator is shifted to $1/(x-x_j(t_i))$—illustrated in Fig. 4.13.

The expressions from each term in Eq. (4.48) are identical to that from the corresponding MHV diagram except for the shifted propagators. Furthermore, each MHV diagram appears twice in the sum; once for each propagator. For the diagram with channels $(x-x_i)$ and $(x-x_j)$ then summing the two contributing terms in the recursion relation we find

$$\int d^4\eta\, d^4\eta'\, \frac{\mathcal{A}_{\text{MHV}}(\{\lambda,\eta\},i,\ldots,j-1,\{\lambda',\eta'\})\mathcal{A}_{\text{MHV}}(\{\lambda',\eta'\},j\ldots,i-1,\{\lambda,\eta\})}{(x-x_i)^2(x-x_j)^2} \tag{4.50}$$

multiplied by an overall factor

$$\left[\frac{(x-x_i)^2}{(x-x_i(z_j))^2}+\frac{(x-x_j)^2}{(x-x_j(z_i))^2}\right]. \tag{4.51}$$

We now employ the following contour integral

$$\oint\frac{dz}{z}\frac{1}{(x-x_i(z))^2(x-x_j(z))^2}=0 \tag{4.52}$$

and find that the terms in the expression (4.51) sum to unity. Hence the recursion relation leads to precisely the MHV diagram expansion for the loop integrand

$$\sum_{1\le i<j\le n}\int d^4\eta\, d^4\eta'\, \frac{\mathcal{A}_{\text{MHV}}(\{\lambda,\eta\},i,\ldots,j-1,\{\lambda',\eta'\})\mathcal{A}_{\text{MHV}}(\{\lambda',\eta'\},j\ldots,i-1,\{\lambda,\eta\})}{(x-x_i)^2(x-x_j)^2}\,. \tag{4.53}$$

as illustrated in Fig. 4.14.

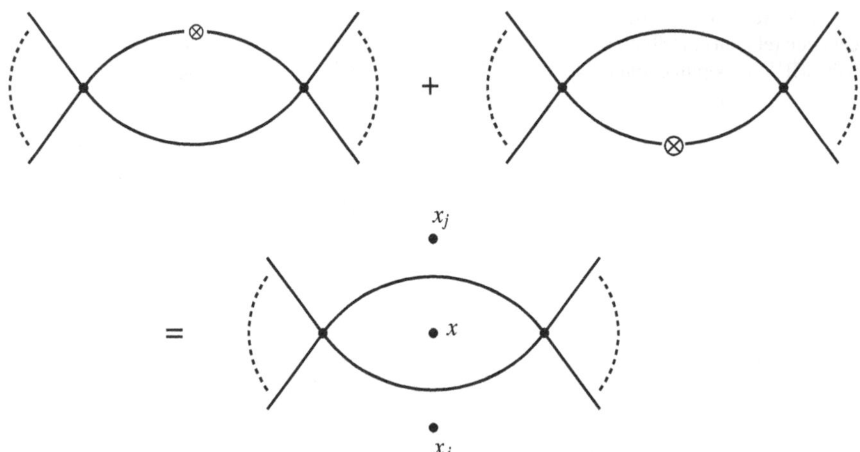

Fig. 4.14 The two terms in the recursion summing to an MHV diagram. Tensor product symbols represent propagators shifted and evaluated on the pole of the other

Let us briefly consider the loop superamplitude itself, ignoring issues of regularization. We first introduce an additional integration variable

$$\frac{\mathrm{d}^4 x}{(x-x_i)^2(x-x_j)^2} = \frac{\mathrm{d}^4 x_1}{(x_1-x_i)^2}\frac{\mathrm{d}^4 x_2}{(x_2-x_j)^2}\delta^4(x_1-x_2) \tag{4.54}$$

and decompose each of them using the reference spinor

$$(x_1-x_i) = \ell_i + z_i q \qquad (x_2-x_j) = \ell_j + z_j q \tag{4.55}$$

where the null reference momentum is $q^{\alpha\dot{\alpha}} = \bar{\zeta}^{\alpha}\zeta^{\dot{\alpha}}$ and the null four-momenta (ℓ_i, ℓ_j) and the parameters (z_i, z_j) are defined by, for example,

$$|\ell_i\rangle = (x_1-x_i)|\zeta] \qquad z_i = \frac{(x_1-x_i)^2}{2q\cdot(x_1-x_i)}. \tag{4.56}$$

The integration measure may then be rewritten in the new variables

$$\begin{aligned}\int \frac{\mathrm{d}^4 x_1}{(x_1-x_i)^2}\frac{\mathrm{d}^4 x_2}{(x_2-x_i)^2}\delta^4(x_1-x_2) &= \int \frac{\mathrm{d}z_1}{z_1}\frac{\mathrm{d}z_2}{z_2}\mathrm{D}^3\ell_1\mathrm{D}^3\ell_2\,\delta^4(x_{ij} \\ &\qquad + \ell_i - \ell_j + (z_i-z_j)q) \\ &= \int \frac{\mathrm{d}z}{z}\mathrm{dLIPS}(\ell_i, -\ell_j, x_{ij}(z)) \end{aligned} \tag{4.57}$$

where $z = (z_i - z_j)$ and $x_{ij}(z) = x_{ij} - zq$. Therefore expanding the NMHV vertex has turned the forward limit, which is a dispersion integral of a single-cut, into a sum of dispersion integrals over standard unitarity cuts. These dispersion integrals reconstruct the standard box expansion from its unitarity cuts as shown in [18].

Momentum Twistor Space

We now perform the same calculation in momentum twistor space. The only terms are forward limits of tree-level NMHV superamplitudes and hence we have

$$\mathcal{A}^{(1)}_{n,0}(1,\ldots,n) = \sum_{i=1}^{n} \oint_{\Gamma} [\,*\,,i-1,i,a,b]\,\mathcal{A}^{(0)}_{n+2,1}(i,\ldots,i-1,a,b';z_i) \quad (4.58)$$

where the contour Γ is chosen so that $\mathcal{Z}_a$ and $\mathcal{Z}_b$ are sent to the intersection point $\mathcal{Z}'_b$. Now expand the tree-level NMHV superamplitude using the MHV diagram expansion and note that terms involving forward limits of adjacent legs on the same vertex are independent of either χ_A or χ_B or both, and hence vanish upon fermionic integration. The remaining terms involve additional channels $(x - x_j)$ and the integrand becomes

$$\sum_{i=1}^{n}\sum_{j\neq i} \frac{(x-x_j)^2}{(x-x_j(z_i))^2} \oint_{\Gamma} [\,*\,,i-1,i,a,b]\,[\,*\,,j-1,j,a,b']\,. \quad (4.59)$$

The all-line deformation of momentum twistors only effects the denominator factor $\langle a,b,j–1,j\rangle$ in the dual superconformal bracket $[\,*\,,j–1,j,a,b']$. This corresponds to the momentum space propagator $1/(x-x_j)^2$ and the effect of the shift has been pulled outside the contour integral.

We now perform the contour integral implementing the forward limit. The contour is chosen to enclose poles in the integrand arising from factors of $\langle a,i-1,i,\,*\,\rangle$ and $\langle b\;j-1,j\,*\rangle$ in the denominator and therefore we have

$$\oint_{S^1\times S^1} \frac{\langle\lambda_a d\lambda_a\rangle\langle\lambda_b d\lambda_b\rangle}{\langle\lambda_a\lambda_b\rangle\langle a,i-1,i,\,*\,\rangle\langle b,j-1,j\,*\rangle} = \frac{1}{\langle\,*\,,i-1,i,[a\,\rangle\langle\,b],j-1,j,\,*\,\rangle}\,. \quad (4.60)$$

The result may be absorbed by introducing the new intersection point $b'' = (a,b)\cap(j-1\;j\,*)$ and replacing the unshifted bracket by $[\,*\,,i-1,i,a,b'']$ so that we now have

$$\sum_{i-1}^{n}\sum_{j\neq i} \frac{(x-x_j)^2}{(x-x_j(z_i))^2}\,[\,*\,,i-1,i,a,b'']\,[\,*\,,i-1,i,a,b']\,. \quad (4.61)$$

Each allowed MHV diagram appears twice in Eq. (4.61) and we use the same residue theorem as in momentum space to combine terms in pairs. The result for the momentum twistor integrand is then

$$\mathcal{A}_{n,0}^{(1)}(1,\ldots,n) = \sum_{1\leq i<j\leq n} [*,i-1,i,a,b''] \, [*,i-1,i,a,b'] \,, \tag{4.62}$$

which is the correct expression.

4.4 Solution of All-Line Recursion

In this section, we prove that the complete solution of the all-line recursion relation is precisely the MHV diagram formalism for all tree-level superamplitudes and loop integrands. The argument extends that presented in [12] for tree-level superamplitudes and proceeds by induction on the grassmann degree and number loops.

4.4.1 Outline of Proof

We first present an outline of the proof.

1. The induction starts with the tree-level NMHV superamplitude. We have shown that the all-line recursion directly generates the MHV expansion of this amplitude.
2. To begin the inductive step, we assume that the MHV diagram expansion correctly reproduces the integrand of all m-loop N^{q}MHV superamplitudes with
   - $q < k$ and $m \leq \ell$
   - $q = k+1$ with $m = \ell - 1$.
3. To complete the inductive step, we consider the all-line recursion relation for the integrand of the ℓ-loop N^{k}MHV superamplitude. This recursion relation involves two kinds of terms:
   - Factorization terms involving m-loop N^{q}MHV integrands with $m \leq \ell$ and $q < k$.
   - Forward terms involving the $(\ell - 1)$-loop N^{k+1}MHV integrand.

   According to our assumption, these terms may be expanded in MHV diagrams. We then show that the all-line recursion relation correctly reproduces the MHV diagram expansion of the ℓ-loop N^{k}MHV integrand.

We will consider the factorization and forward terms separately.

4.4.2 Factorization Terms

Consider the recursion relation for the ℓ-loop N^{k}MHV superamplitude.

We first examine the factorization terms from the channel I with four-momentum $P_I = (x_i - x_j)$ bounding two external regions. The contribution to the recursion relation from terms with N^{q}MHV and N^{k-q+1}MHV superamplitudes with m and $(\ell - m)$ loops respectively is

$$\int \mathrm{d}^4\eta_I \, \mathcal{A}_q^{(m)}(i, \ldots, j-1, I; z_I) \, \frac{1}{P_I^2} \, \mathcal{A}_{k-q-1}^{(\ell-m)}(j, \ldots, i-1, -I, z_I) \,. \tag{4.63}$$

These terms must be summed over the degrees of the integrands $(0 \leq q \leq k-1)$, the numbers of loops $(0 \leq m \leq \ell)$ and symmetrized over the assignment of all internal region momenta. All factorization channels are then obtained by summing over the range $(1 \leq i < j \leq n)$. The terms where i and j are separated by fewer than two (modulo n) will vanish automatically.

We now replace the subintegrands with their MHV diagram expansions using the same reference spinor $\zeta^{\dot{\alpha}}$. Each term in the expansions of the sub-integrands then depends on the shifted internal momentum $P_I(t_I) = \lambda_I \tilde{\lambda}_I$ through the holomorphic CSW spinor $\lambda_I = P_I|\zeta]$ only. Hence each term in the expansion corresponds to an MHV diagram for the original ℓ-loop N^{k}MHV integrand except that any propagators bounding external regions (except $1/P_I^2$) are shifted and evaluated on the pole z_I where $P_I(t)$ becomes null.

Summing over degrees $0 \leq q \leq k-1$, numbers of loops $0 \leq m \leq \ell$ and symmetrizing over all internal regions, each MHV diagram for the ℓ-loop N^{k}MHV containing the channel I appears once in the expansion of the factorization channel I.

4.4.3 Forward Terms

Forward terms in the recursion relation arise from channels I with momentum $P_I = (x - x_i)$ bounding an external and an internal region. These terms involve the forward limit of the N^{k+1}MHV integrand with $(l-1)$ loops and have the following contribution to the recursion relation

$$\frac{1}{P_I^2} \int \mathrm{d}^4\eta_I \, \mathcal{A}_{n+2,k+1}^{(\ell-1)}(I, -I, i, \ldots, i-1; z_I) \,. \tag{4.64}$$

which must be symmetrized over the internal region momenta in the forward channel. All forward terms are found by summing over $i = 1, \ldots, n$.

We expand the $(\ell - 1)$-loop N^{k+1}MHV integrand using the MHV expansion. Terms involving forward limits of adjacent legs on the same vertex vanish on fermionic integration. The remaining terms depend on $P_I(z_I)$ through the holomorphic spinor $\lambda_I = P_I |\zeta]$ and hence correspond to MHV diagrams where all propagators bounding external regions, except for $1/P_I^2$, are shifted and evaluated on the pole z_I where $P_I(t)$ becomes null.

Symmetrizing over the choice of region coordinates to the forward channels, then each MHV diagram for the ℓ-loop N^{k}MHV integrand containing the channel I appears exactly once in the expansion of the forward channel I.

4.4.4 Completion of Proof

Consider now the particular MHV diagram for the ℓ-loop N^{k}MHV integrand with channels

$$\{I_1, \ldots, I_P\} \tag{4.65}$$

where $P = k + 2\ell$ is the number of propagators in each MHV diagram. Suppose that only $P' \leq P$ of the channels are bounding external regions and are therefore affected by the all-line deformation. We choose to order the channels so that those bounding external regions appear first

$$\{I_1, \ldots, I_{P'}, \ldots, I_P\}\,. \tag{4.66}$$

This particular diagram occurs exactly P' times in the recursion relation—once for each channel $\{I_1, \ldots, I_{P'}\}$ bounding an external region. Each term contributes an expression equal to the MHV diagram, except that $(P' - 1)$ of the propagators are deformed and evaluated on the pole of the remaining one. The dependence on the shifted propagators may be factored out of each term, leaving the correct contribution from the MHV diagram, multiplied by a ratio of propagators. For example, the term arising from the channel I_1 contributes

$$\mathcal{A}_{n,k}^{(\ell)}(I_1, \ldots, I_P)\,\frac{P_{I_2}^2 \ldots P_{I_{P'}}^2}{P_{I_2}^2(z_{I_1}) \ldots P_{I_{P'}}^2(z_{I_1})}\,, \tag{4.67}$$

where we have denoted the contribution from the MHV diagram containing the channels $\{I_1, \ldots, I_P\}$ by $A_{n,k}^{(0)}(I_1, \ldots, I_P)$. Summing over all of the terms, we find the following expression

$$\mathcal{A}_{n,k}^{(\ell)}(I_1, \ldots, I_P)\sum_{j=1}^{P'} \frac{P_{I_1}^2 \ldots P_{I_{P'}}^2}{P_{I_1}^2(z_{I_j}) \ldots P_{I_{j-1}}^2(z_{I_j}) P_{I_j}^2 P_{I_{j+1}}^2(z_{I_j}) \ldots P_{I_{P'}}^2(z_{I_j})}\,. \tag{4.68}$$

However, using the result of the contour integral

$$\oint \frac{dz}{z} \frac{P_{I_1}^2 \dots P_{I_{P'}}^2}{P_{I_1}^2(z) \dots P_{I_{P'}}^2(z)} = 0 \tag{4.69}$$

the summation collapses to unity and we recover only the contribution from the MHV diagram

$$\mathcal{A}_{n,k}^{(k)}(I_1, \dots, I_P) \, . \tag{4.70}$$

Now running over all of the allowed MHV diagrams for the ℓ-loop N^{k}MHV integrand we pick up each term in the recursion relation exactly once. Hence the recursion relation has generated the MHV diagram expansion for the ℓ-loop N^{k}MHV integrand. This completes the induction.

References

1. R. Britto, F. Cachazo, B. Feng, New recursion relations for tree amplitudes of gluons. Nucl. Phys. **715**, 499–522 (2005, hep-th/0412308)
2. R. Britto, F. Cachazo, B. Feng, E. Witten, Direct proof of tree-level recursion relation in Yang-Mills theory. Phys. Rev. Lett. **94**, 181602 (2005, hep-th/0501052)
3. S. Badger, E. Glover, V. Khoze, P. Svrcek, Recursion relations for gauge theory amplitudes with massive particles. JHEP **0507**, 025 (2005, hep-th/0504159)
4. S. Badger, E. Glover, V.V. Khoze, Recursion relations for gauge theory amplitudes with massive vector bosons and fermions. JHEP **0601**, 066 (2006, hep-th/0507161)
5. Z. Bern, L.J. Dixon, D.A. Kosower, On-shell recurrence relations for one-loop QCD amplitudes. Phys. Rev. **D71**, 105013 (2005, hep-th/0501240)
6. Z. Bern, L.J. Dixon, D.A. Kosower, The last of the finite loop amplitudes in QCD. Phys. Rev. **D72**, 125003 (2005, hep-ph/0505055)
7. A. Brandhuber, P. Heslop, G. Travaglini, A note on dual superconformal symmetry of the N=4 super Yang-Mills S-matrix. Phys. Rev. **D78**, 125005 (2008, arXiv:0807.4097)
8. N. Arkani-Hamed, F. Cachazo, J. Kaplan, What is the simplest quantum field theory? JHEP. **1009**, 016 (2010, arXiv:0808.1446)
9. N. Arkani-Hamed, J.L. Bourjaily, F. Cachazo, S. Caron-Huot, J. Trnka, The all-loop integrand for scattering amplitudes in planar N=4 SYM. JHEP **1101**, 041 (2011, arXiv:1008.2958)
10. R. H. Boels, On BCFW shifts of integrands and integrals. JHEP **1011**, 113 (2010, arXiv:1008.3101)
11. H. Elvang, D.Z. Freedman, M. Kiermaier, Recursion relations, generating functions, and unitarity sums in N=4 SYM theory. JHEP **0904**, 009 (2009, arXiv:0808.1720)
12. H. Elvang, D.Z. Freedman, M. Kiermaier, Proof of the MHV vertex expansion for all tree amplitudes in N=4 SYM theory. JHEP **0906**, 068 (2009, arXiv:0811.3624)
13. M. Bullimore, MHV diagrams from an all-line recursion relation, arXiv:1010.5921
14. J. Drummond, J. Henn, All tree-level amplitudes in N=4 SYM. JHEP **0904**, 018 (2009, arXiv:0808.2475)
15. T. Cohen, H. Elvang, M. Kiermaier, On-shell constructibility of tree amplitudes in general field theories. JHEP **1104**, 053 (2011, arXiv:1010.0257)
16. F. Cachazo, P. Svrcek, E. Witten, MHV vertices and tree amplitudes in gauge theory. JHEP **0409**, 006 (2004, hep-th/0403047)
17. M. Bullimore, L. Mason, D. Skinner, MHV diagrams in momentum twistor space. JHEP **1012**, 032 (2010, arXiv:1009.1854)
18. A. Brandhuber, B.J. Spence, G. Travaglini, One-loop gauge theory amplitudes in N=4 super Yang-Mills from MHV vertices. Nucl. Phys. **B706**, 150–180 (2005, hep-th/0407214)

Chapter 5
Wilson Loops

In the previous chapter, we have seen that the kinematics of scattering amplitudes in planar $\mathcal{N} = 4$ super Yang-Mills theory are encoded in the momentum twistor polygon

$$C = X_1 \cup X_2 \cup \cdots X_n \,, \tag{5.1}$$

which is determined by the momentum twistors $\{\mathcal{Z}_1, \ldots, \mathcal{Z}_n\}$. This is a holomorphic curve in $\mathbb{CP}^{3|4}$ and in fact a degenerate elliptic curve. We can now ask the question–is there a natural observable $\mathcal{O}_C$ depending on the curve C whose quantum expectation value computes the all planar scattering amplitudes (Fig. 5.1)?

An important clue to finding the correct observable $\mathcal{O}_C$ comes from the form of the action in twistor space. In Chap. 2, we explained how self-dual sector of maximally supersymmetric Yang-Mills theory is described by holomorphic Chern-Simons theory in twistor space

$$S_1(\mathcal{A}) = \int \mathrm{D}^{3|4}Z \wedge \mathrm{tr}\left(\mathcal{A} \wedge \bar{\partial}\mathcal{A} + \frac{2}{3}\mathcal{A} \wedge \mathcal{A} \wedge \mathcal{A}\right). \tag{5.2}$$

whose equations of motion impose the twistor bundle $E \to U \subset \mathbb{CP}^{3|4}$ be holomorphic. The partial connection $\mathcal{A}$ is a (0, 1)-form valued in the Lie algebra of the gauge group. In addition, there is a unique meromorphic (1, 0)-form ω on the nodal curve C whose only singularities are simple poles at the nodes with residue ± 1. Hence we can construct a gauge invariant operator

$$\mathrm{Tr}\,\mathrm{P}\exp \int_C \omega \wedge \mathcal{A}. \tag{5.3}$$

that is the holomorhic analogue of a Wilson loop. Due to the close connection between determinants and holonomy, the holomorphic Wilson loop remains a natural observable even when we include the interaction terms $S_2(\mathcal{A})$ that expand around the self-dual sector.

M. R. Bullimore, *Scattering Amplitudes and Wilson Loops in Twistor Space*, Springer Theses, DOI: 10.1007/978-3-319-00909-4_5,

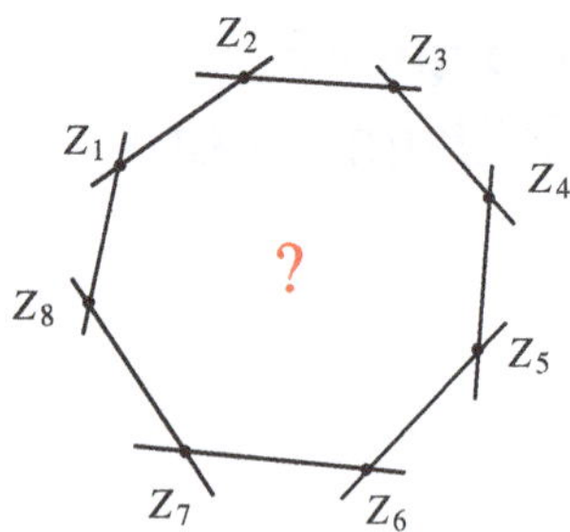

Fig. 5.1 In this chapter we ask which observable computes the planar $\mathcal{S}$-matrix in momentum twistor space

In this section, we construct the holomorphic Wilson loop and present evidence that it computes planar tree-level superamplitudes and loop integrands, following reference [1] by Mason and Skinner. Then we derive a holomorphic analogue of the 'loop equation' [2], describing how the quantum expectation value changes when the nodal curve $\mathcal{C}$ is deformed holomorphically. For particular deformations, the loop equations coincide with the BCFW or all-line recursion relations. Since planar tree-level superamplitudes and loop integrands are uniquely determined by such recursion relations, this proves the duality [3].

5.1 The Holomorphic Wilson Loop

In this chapter, we will think of the momentum twistor polygon more abstractly as a nodal Riemann surface

$$\Sigma = \Sigma_1 \cup \Sigma_2 \cup \cdots \cup \Sigma_n \tag{5.4}$$

where each component of the surface is a Riemann sphere $\Sigma_j \cong \mathbb{CP}^1$, together with a holomorphic map into twistor space

$$\begin{aligned} \mathcal{Z} : \Sigma &\rightarrow \mathbb{CP}^{3|4} \\ : \sigma &\rightarrow \mathcal{Z}(\sigma) \end{aligned} \tag{5.5}$$

which is linear on each component and sends $\mathcal{Z} : \Sigma_j \rightarrow X_j$. The components Σ_j each have two marked points at the intersection points $\sigma_{j-1} = \Sigma_{j-1} \cap \Sigma_j$ and $\sigma_j = \Sigma_j \cap \Sigma_{i+1}$, which are sent to the corresponding momentum twistors $\mathcal{Z}(\sigma_j) = \mathcal{Z}_j$.

5.1.1 Some Definitions

The starting point for constructing the holomorphic Wilson loop operator is the holomorphic frame $h(\sigma)$ on each component Σ_j. This provides a holomorphic trivialization for the pull-back of the twistor bundle $E \to U \subset \mathbb{CP}^{3|4}$ to the component Σ_j. Let us introduce the notation $\mathcal{A}(\sigma) = (\mathcal{Z}^*\mathcal{A})(\sigma)$ for the pull-back of the partial connection so that the holomorphic frame obeys

$$\big(\bar{\partial}_\sigma + \mathcal{A}(\sigma)\big)h = 0. \tag{5.6}$$

This equation only defines the holomorphic frame up to gauge transformations $h(\sigma) \to h(\sigma)g$ where g is globally holomorphic in σ and hence constant. In particular, we can choose the following holomorphic frame

$$\mathrm{U}(\sigma, \sigma_0) \equiv h(\sigma)h^{-1}(\sigma_0) \tag{5.7}$$

to be the particular solution of Eq. (5.6) satisfying the boundary condition that $\mathrm{U}(\sigma_0, \sigma_0) = \mathbb{I}$. The holomorphic frame then defines a holomorphic map between fibres of the bundle at the points σ and σ_0 on the component Σ_j. Thus it is a holomorphic analogue of the parallel propagator.

It follows immediately from the definition (5.7) that the holomorphic parallel propagator obeys the concatenation and inversion properties

$$\mathrm{U}(\sigma_2, \sigma_1)\mathrm{U}(\sigma_1, \sigma_0) = \mathrm{U}(\sigma_2, \sigma_0) \tag{5.8}$$

$$\mathrm{U}(\sigma_0, \sigma) = \mathrm{U}(\sigma, \sigma_0)^{-1}. \tag{5.9}$$

Similarly, under a smooth gauge transformation $h(\sigma) \to h(\sigma)g(\sigma)$

$$\mathrm{U}(\sigma, \sigma_0) \to g(\sigma)\mathrm{U}(\sigma, \sigma_0)g^{-1}(\sigma_0). \tag{5.10}$$

These properties are the direct analogues of those obeyed by the parallel propagator between points on a real curve.

The definitions in Eqs. (5.6) and (5.7) completely determine the holomorphic parallel propagator. However, for concrete calculations it is necessary to have an explicit perturbative expansion of the solution in powers of $\mathcal{A}(\sigma)$. We now consider the abelian and non-abelian theories in turn.

5.1.2 Abelian Theory

In the abelian theory, the twistor bundle $E \to U \subset \mathbb{CP}^{3|4}$ is simply a line bundle. Thus the holomorphic frame $h(\sigma)$ on any component Σ_j is a smooth function, and we can set $U(\sigma, \sigma_0) = \exp(\phi(\sigma))$ for some smooth function $\phi(\sigma)$. Then the holomorphic propagator is determined by the equation

$$\bar{\partial}\,\phi(\sigma) + \mathcal{A}(\sigma) = 0 \tag{5.11}$$

with the boundary condition $\phi(\sigma_0) = 0$.

The solution to this problem is

$$\phi(\sigma) = -\int \omega(\sigma, \sigma') \wedge \mathcal{A}(\sigma') \tag{5.12}$$

where

$$\omega(\sigma, \sigma') = \frac{(\sigma - \sigma_0)}{(\sigma - \sigma')(\sigma' - \sigma_0)}\, d\sigma' \tag{5.13}$$

is the unique meromorhic form on the Riemann sphere Σ_j whose only singularities are simple poles at $\sigma' = \sigma$ and $\sigma' = \sigma_0$ with residues ± 1. This is precisely the Green's function for the $\bar{\partial}$-operator on the Riemann sphere with the appropriate boundary conditions.

Thus the parallel propagator can be expressed as

$$\mathrm{U}(\sigma, \sigma_0) = \exp\left(-\int \omega \wedge \mathcal{A}\right) \tag{5.14}$$

where it is understood that the Green's function has the appropriate boundary conditions for the arguments.

5.1.3 Non-Abelian Theory

In the non-abelian theory, the solution can be expanded in a series solution by formally inverting the covariant $\bar{\partial}$-operator, taking into account the boundary condition,

$$\begin{aligned} \mathrm{U}(\sigma, \sigma_0) &= 1 + \sum_{m=1}^{\infty} (-1)^m (\bar{\partial}^{-1} \mathcal{A}(\sigma))^m \\ &= 1 + \sum_{m=1}^{\infty} (-1)^m \int_{X^m} \bigwedge_{i=1}^{m} \left(\omega(\sigma_i) \wedge \mathcal{A}(\sigma_i)\right). \end{aligned} \tag{5.15}$$

In the first line all $\bar{\partial}^{-1}$ operators are understood to act on everything to their right. This means that the Green's functions $\omega(\sigma_i)$ in the second line have simple poles at the starting point σ_0 and at the next point σ_{i+1} in the colour ordering (or at σ when $i = m$). Thus, completely explicitly, we have

$$
\begin{aligned}
\mathrm{U}(\sigma, \sigma_0) &= 1 + \sum_{m=1}^{\infty} (-1)^m \int_{X^m} \frac{(\sigma - \sigma_0)\, d\sigma_1 \ldots d\sigma_m}{(\sigma - \sigma_m) \ldots (\sigma_2 - \sigma_1)(\sigma_1 - \sigma_0)} \mathcal{A}(\sigma_m) \ldots \mathcal{A}(\sigma_1) \\
&\equiv \mathrm{P} \exp\left(- \int \omega \wedge \mathcal{A} \right), \qquad (5.16)
\end{aligned}
$$

where in the final line we have introduced the holomorphic 'path ordering' symbol for the perturbative expansion.

Now we can compose the holomorphic parallel propagators from each component Σ_j, holomorphically transporting from one node to the next, and then take the trace in the fundamental representation

$$
\begin{aligned}
\mathrm{W}(\mathcal{Z}_1, \ldots, \mathcal{Z}_n) &= \mathrm{Tr}\left(\mathrm{U}_{\Sigma_2}(\sigma_1, \sigma_2)\, \mathrm{U}_{\Sigma_3}(\sigma_2, \sigma_3) \ldots \mathrm{U}_{\Sigma_1}(\sigma_n, \sigma_1) \right) \\
&= \mathrm{Tr}\, \mathrm{P} \exp\left(- \int \omega \wedge \mathcal{A} \right). \qquad (5.17)
\end{aligned}
$$

The dependence on the momentum twistors $\{\mathcal{Z}_1, \ldots, \mathcal{Z}_n\}$ is through the pull-back of the partial connection by the map $\mathcal{Z} : \Sigma \to \mathbb{CP}^{3|4}$. This is the holomorphic Wilson loop that we consider in the following sections.

5.2 Amplitude/Wilson Loop Duality

We now consider the expectation value of the holomorphic Wilson loop in twistor space, using the twistor action for maximally supersymmetric Yang-Mills theory,

$$
\langle \mathrm{W}(\mathcal{Z}_1, \ldots, \mathcal{Z}_n) \rangle = \int \mathcal{D}\mathcal{A}\, \mathrm{W}(\mathcal{Z}_1, \ldots, \mathcal{Z}_n)\, \exp\left(-S(\mathcal{A})\right). \qquad (5.18)
$$

The action is invariant under gauge transformations

$$
\mathcal{A} \to \mathcal{A} + \bar{\partial}\epsilon + [\mathcal{A}, \epsilon], \qquad (5.19)
$$

and hence in order to make sense of the path integral we have implement the Fadeev-Popov procedure, introducing gauge fixing and ghost contributions. This can be done and the details can be found in [1]. Here I will consider only an axial gauge, where the ghosts decouple.

Let us first consider the expectation value in the holomorphic Chern-Simons theory action $S_1(\mathcal{A})$, corresponding to the self-dual sector of $\mathcal{N} = 4$ super Yang-Mills theory. It is conjectured that is that this expectation value computes the complete planar $\mathcal{S}$-matrix at tree-level,

$$
\langle \mathrm{W}(\mathcal{Z}_1, \ldots, \mathcal{Z}_n) \rangle = \mathcal{A}_n^{(0)}(\mathcal{Z}_1, \ldots, \mathcal{Z}_n). \qquad (5.20)
$$

The complete twistor action involves a perturbative expansion around the self-dual sector of the theory. The interaction term

$$\begin{aligned} S_2(\mathcal{A}) &= \int \mathrm{d}^{4|8}X\ \mathcal{L}(X) \\ &= \int \mathrm{d}^{4|8}X\ \log\det(\bar{\partial} + \mathcal{A})|_X \end{aligned} \tag{5.21}$$

involves an integral over lines in twistor space, or equivalently, over chiral superspace. Thus the perturbative expansion of the holomorphic Wilson loop expectation value can be computed by the method of lagrangian insertions,

$$\sum_{l=0}^{\infty} g^{2l} \int \mathrm{d}^{4|8}X_1 \dots \mathrm{d}^{4|8}X_l \left\langle \mathrm{W}(\mathcal{Z}_1, \dots, \mathcal{Z}_n)\, \mathcal{L}(X_1) \dots \mathcal{L}(X_l) \right\rangle \tag{5.22}$$

where the expectation value is taken in holomorphic Chern-Simons theory, or equivalently, in self-dual maximally supersymmetric Yang-Mills theory.

Now stripping away the chiral superspace integrals, we obtain a canonical notion of integrand for the holomorphic Wilson loop expectation value. The conjecture is that the holomorphic Wilson loop integrand coincides with the superamplitude integrand, as follows

$$\left\langle \mathrm{W}(\mathcal{Z}_1, \dots, \mathcal{Z}_n)\, \mathcal{L}(X_1) \dots \mathcal{L}(X_l) \right\rangle = \mathcal{A}_n^{(l)}(\mathcal{Z}_1, \dots, \mathcal{Z}_n; X_1, \dots, X_l). \tag{5.23}$$

where the correlator is taken in holomorphic Chern-Simons theory.

We emphasize that this is a statement about four-dimensional integrands. Integrating over the real Minkowski slice $\mathbb{M}^{4|8}$ leads to divergences from regions of integration where the line X intersections the momentum twistor polygon. This requires the introduction of a regulation scheme, for example, dimensional regularization. One might conjecture that the regulated and integrated objects are equal in some regularization scheme. This stronger statement leads to many subtle issues that we do not attempt to address here [4, 5].

5.3 Perturbative Checks

In this section, we evaluate some simple examples of the correlation functions

$$\left\langle \mathrm{W}(\mathcal{Z}_1, \dots, \mathcal{Z}_n)\mathcal{L}(X_1) \dots \mathcal{L}(X_\ell) \right\rangle \tag{5.24}$$

in holomorphic Chern-Simons theory on twistor space. In axial gauge, we find that Feynman diagrams for the above correlation function are in one-to-one correspondence with MHV diagrams for the superamplitude integrand. The calculations were

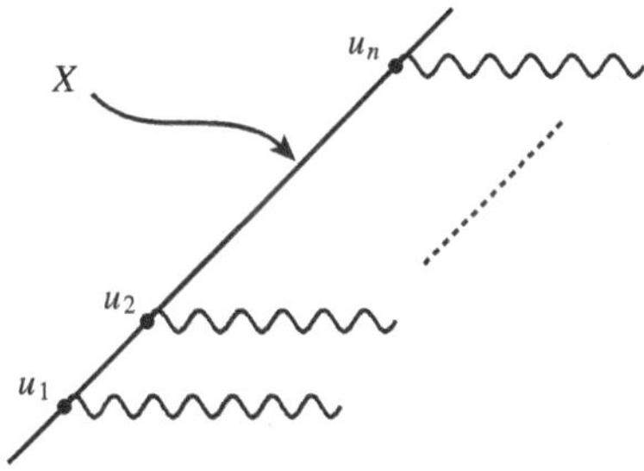

Fig. 5.2 Expanding the lagrangian insertion leads to an infinite series of non-local interaction terms in twistor space

performed by Mason and Skinner [1] and we simply review their work here for reasons of completeness.

5.3.1 Preliminaries

In the following computations, we choose the map $\mathcal{Z} : \Sigma \to \mathbb{CP}^{3|4}$ whose restriction to each component of the Riemann surface Σ_j is given by

$$\mathcal{Z}_j(\sigma) = \mathcal{Z}_{j-1} + \sigma \mathcal{Z}_j. \tag{5.25}$$

The perturbative expansion of the holomorphic parallel propagator derived in then takes the following form

$$\mathrm{U}_{X_j}(\mathcal{Z}_j, \mathcal{Z}_{j-1}) = 1 + \sum_{m=1}^{\infty} (-1)^m \int_{X_j^m} \frac{\mathrm{d}\sigma_m \cdots \mathrm{d}\sigma_1}{(\sigma_m - \sigma_{m-1}) \cdots (\sigma_2 - \sigma_1)\sigma_1} \mathcal{A}(\sigma_m) \cdots \mathcal{A}(\sigma_1). \tag{5.26}$$

Similarly, we represent the twistor line associated to each lagrangian insertion $\mathcal{L}(X)$ by a linear map from the Riemann sphere into twistor space $\mathcal{Z}(\rho) = \mathcal{Z}_a + \rho\, \mathcal{Z}_b$. Then the perturbative expansion of the lagrangian insertion becomes

$$\mathcal{L}(X) = \sum_{n=2}^{\infty} \frac{(-1)^n}{n} \int_{X^n} \frac{\mathrm{d}\rho_n \cdots \mathrm{d}\rho_1}{(\rho_1 - \rho_n) \cdots (\rho_2 - \rho_1)(\rho_1 - \rho_n)} \operatorname{Tr} \mathcal{A}(\rho_n) \cdots \mathcal{A}(\rho_1). \tag{5.27}$$

Thus, the lagrangian insertions lead to an infinite tower of interaction terms consisting of $n \geq 2$ insertions of the twistor partial connection $\mathcal{A}$ integrated over the internal line X (Fig. 5.2).

To compute the expectation values we have choose a gauge and find the corresponding propagator. Here we will impose an axial gauge condition

$$\overline{V} \lrcorner \mathcal{A} = 0, \tag{5.28}$$

where $\mathcal{V}$ is any holomorphic vector field on twistor space and $\overline{\mathcal{V}}$ denotes the conjugate. The simplest choice is to pick a reference twistor $\mathcal{Z}_*$ and then define the holomorphic vector field $\mathcal{V} = \mathcal{Z}_*^I\, \partial/\partial \mathcal{Z}^I$. The axial gauge condition then simply requires that $\mathcal{A}$ vanishes on restriction to lines through the reference point.

An advantage of axial gauge is that ghosts excitations decouple, and hence we are free to use holomorphic Chern-Simons theory alone. Furthermore, there are now only two independent components of the partial connection $\mathcal{A}$, so the cubic term vanishes automatically leaving only the kinetic term

$$\int \mathrm{D}^{3|4}\mathcal{Z} \wedge \mathrm{Tr}\big(\mathcal{A} \wedge \bar{\partial}\mathcal{A}\big). \tag{5.29}$$

Thus, the axial gauge propagator obeys

$$\bar{\partial}\, \Delta(\mathcal{Z}_1, \mathcal{Z}_2) = \bar{\delta}^{3|4}(\mathcal{Z}_1, \mathcal{Z}_2)\,, \tag{5.30}$$

and the solution is

$$\begin{aligned} \Delta(\mathcal{Z}_1, \mathcal{Z}_2) &= \bar{\delta}^{2|4}(\mathcal{Z}_1, \mathcal{Z}_2, \mathcal{Z}_*) \\ &:= \int \frac{D^2 c}{c_1 c_2 c_3} \bar{\delta}^{4|4}(c_1 \mathcal{Z}_1 + c_2 \mathcal{Z}_2 + c_3 \mathcal{Z}_*). \end{aligned} \tag{5.31}$$

which has support when the points $\mathcal{Z}_1$, $\mathcal{Z}_2$ and $\mathcal{Z}_*$ are collinear. The propagator has singularities when any of pair of points approach one another along the common line. The diagonal singularity when $\mathcal{Z}_1 \to \mathcal{Z}_2$ is the physical singularity required for the definition (5.30) but the additional spurious singularities involving the reference point mean that

$$\bar{\partial}\Delta(\mathcal{Z}_1, \mathcal{Z}_2) = \bar{\delta}^{3|4}(\mathcal{Z}_1, \mathcal{Z}_2) + \bar{\delta}^{3|4}(\mathcal{Z}_2, \mathcal{Z}_*) + \bar{\delta}^{3|4}(\mathcal{Z}_*, \mathcal{Z}_1)\,. \tag{5.32}$$

However, such spurious singularities will vanish in perturbative calculations for generic choices of the reference twistor $\mathcal{Z}_*$.

5.3.2 NMHV Tree

Consider first the expectation value of the holomorphic Wilson loop $\langle \mathrm{W}(\mathcal{Z}_1, \ldots, \mathcal{Z}_n)\rangle$ in holomorphic Chern-Simons theory. The zeroth order term in the fermionic expansion is unity. The Feynman diagrams with the next smallest fermionic degree, $\mathcal{O}(\chi^4)$, have a single propagator connecting two components X_i and X_j of the momentum twistor polygon—see Fig. 5.3. Evaluating this Feynman diagram we find

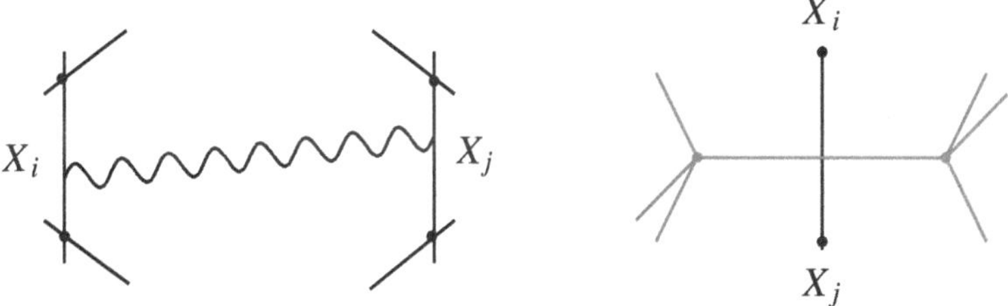

Fig. 5.3 Planar duality between axial gauge Feynman diagrams and MHV diagrams for the NMHV tree-level superamplitude

$$\int \frac{\mathrm{d}\sigma}{\sigma}\frac{\mathrm{d}\sigma'}{\sigma'}\Delta(Z_i(\sigma), Z_j(\sigma')) = \int \frac{\mathrm{d}\sigma}{\sigma}\frac{\mathrm{d}\sigma'}{\sigma'}\frac{D^2 c}{c_1 c_2 c_3}\,\bar{\delta}^{4|4}\big(c_1\mathcal{Z}_* + c_2\mathcal{Z}_i(\sigma) + c_3\mathcal{Z}_j(\sigma')\big)$$
$$= [\,*\,, i-1, i, j-1, j]. \tag{5.33}$$

Hence, each Feynman diagram is precisely equal to the planar dual MHV diagram for the tree-level NMHV superamplitude—see Fig. 5.3. Summing all such diagrams we obtain the complete tree-level NMHV superamplitude.

5.3.3 N²MHV Tree

Feynman diagrams of the next smallest grassmann degree, $\mathcal{O}(\chi^8)$, have two propagators joining separate components of the momentum twistor polygon. For propagators joining completely separated lines X_i to X_j and X_k to X_l, we find immediately from the above calculation that

$$[\,*\,, i-1, i, j-1, j]\,[\,*\,, k-1, k, l-1, l]. \tag{5.34}$$

When two propagators join the components X_i and X_l to the same line X_j, as illustrated in Fig. 5.4, we find that

$$\int \frac{\mathrm{d}\sigma}{\sigma}\frac{\mathrm{d}\rho\,\mathrm{d}\rho'}{\rho'(\rho'-\rho)}\frac{\mathrm{d}\sigma'}{\sigma'}\,\Delta\big(\mathcal{Z}_i(\sigma), \mathcal{Z}_j(\rho)\big)\,\Delta\big(\mathcal{Z}_j(\rho'), \mathcal{Z}_l(\sigma')\big)$$
$$= [\,*\,, i-1, i, j-1, j']\,[\,*\,, j-1, j, l-1, l]. \tag{5.35}$$

where $\mathcal{Z}_j' = \big\langle *, l-1, l, [\mathcal{Z}_{j-1}\big\rangle\mathcal{Z}_j]$. Again, the holomorphic Wilson loop Feynman diagrams are in 1-1 correspondence with the planar dual MHV diagrams for scattering amplitudes. The same calculations can be straightforwardly extended to the whole expectation value $\big\langle \mathrm{W}(\mathcal{Z}_1, \ldots, \mathcal{Z}_n)\big\rangle$ and the complete planar tree-level $\mathcal{S}$-matrix.

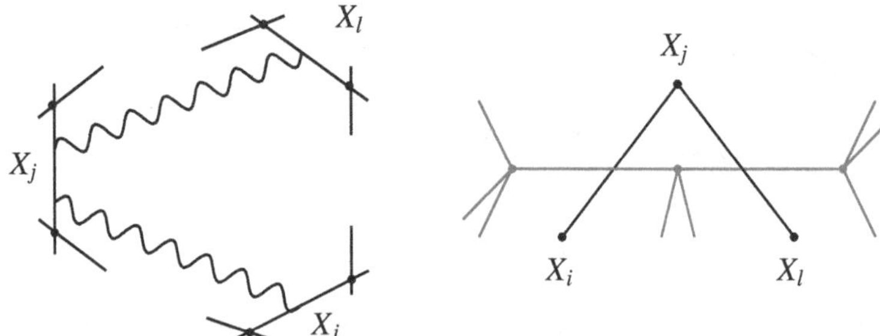

Fig. 5.4 Planar duality between axial gauge Feynman diagrams and MHV diagrams for the N^2 MHV tree-level superamplitude

5.3.4 MHV 1-Loop

Now consider a single lagrangian insertion $\langle \mathrm{W}(\mathcal{Z}_1, \ldots, \mathcal{Z}_n)\, \mathcal{L}(X)\rangle$. Now there are Feynman diagrams with zero grassmann degree, which are obtained by connecting the quadratic term in $\mathcal{L}(X)$ with two separate edges X_i and X_j of the holomorphic Wilson loop, as illustrated in Fig. 5.5. Recalling the parametrization $\mathcal{Z}(u) = \mathcal{Z}_a + \rho\, \mathcal{Z}_b$ of the internal twistor line X, we find

$$\int \frac{\mathrm{d}\sigma}{\sigma} \frac{\mathrm{d}\rho\, \mathrm{d}\rho'}{(\rho' - \rho)^2} \frac{\mathrm{d}\sigma'}{\sigma'}\, \Delta(Z_i(\sigma), Z(\rho))\, \Delta(Z(\rho'), Z_j(\sigma'))$$
$$= [*, i-1, i, a, b']\,[*, j-1, j, a, b''] \tag{5.36}$$

where

$$\mathcal{Z}_b' = \langle *, i-1, i, [\mathcal{Z}_a \rangle\, \mathcal{Z}_b] \qquad \mathcal{Z}_b'' = \big\langle *, i-1, i, [\mathcal{Z}_a \big\rangle\, \mathcal{Z}_b] \tag{5.37}$$

This is once again identical to the contribution to the one-loop superamplitude integrand from an MHV diagram which is planar dual to the momentum twistor Feynman diagram.

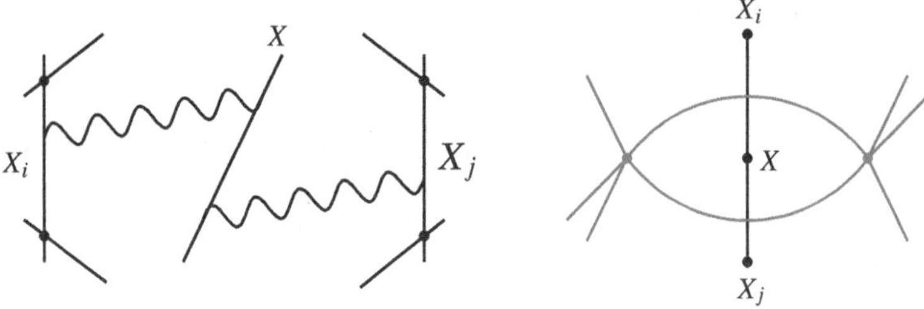

Fig. 5.5 An example of planar duality for the one-loop MHV integrand

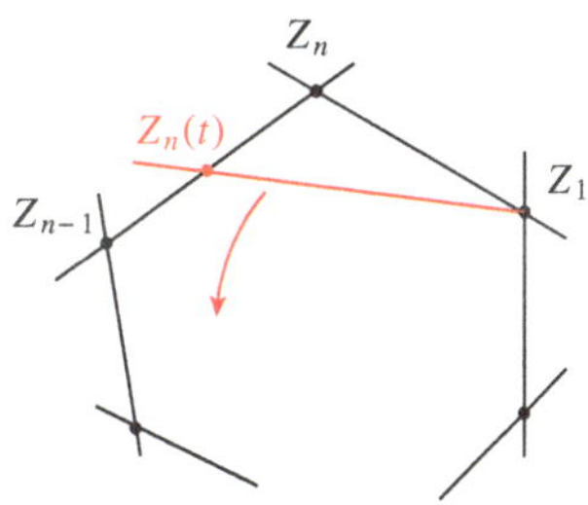

Fig. 5.6 A holomorphic deformation of the momentum twistor polygon

In reference [1], further computations of momentum twistor Feynman diagrams were performed. In all cases, the axial gauge Feynman diagrams for the holomorphic Wilson loop are in one-to-one correspondence with MHV diagrams for the superamplitude integrand. This provided substantial evidence for the duality.

5.4 Holomorphic Loop Equations

In this section, our aim is to further understand and prove the supersymmetric amplitude—Wilson loop duality. Our strategy will be to show that the holomorphic Wilson loop expectation value has the same analytic behaviour as the planar $\mathcal{S}$-matrix, encoded concretely via on-shell recursion relations.

5.4.1 Holomorphic Deformations

We now consider how the expectation value of the holomorphic Wilson loop behaves under holomorphic deformations of the map

$$\mathcal{Z}(t) : \Sigma \longrightarrow \mathbb{CP}^{3|4} . \tag{5.38}$$

Here we need only one-parameter holomorphic deformations labelled by a single complex parameter t, although the extension to more general deformations is straightforward. A simple example is to deform the image of a single node

$$\mathcal{Z}(t) : \sigma_n \longrightarrow \mathcal{Z}_n - t\mathcal{Z}_{n-1} \tag{5.39}$$

corresponding to the BCFW deformation of the momentum twistor polygon. In order to streamline notation, in what follows we will denote the image of the deformed map $\mathcal{Z}(t)$ in twistor space by $C(t)$, that is the deformed momentum twistor polygon labelled by the momentum twistors $\{\mathcal{Z}_1, \ldots, \mathcal{Z}_{n-1}, \mathcal{Z}_n - t\mathcal{Z}_{n-1}\}$.

For any one-parameter holomorphic family maps $\mathcal{Z}(t)$, we will prove that

$$\frac{\partial}{\partial \bar{t}} W(\mathcal{C}(t)) = \int\limits_{\mathcal{C}(t)} \omega(\mathcal{Z}) \wedge d\bar{\mathcal{Z}}^{\bar{I}} \wedge \frac{\partial \bar{\mathcal{Z}}^{\bar{J}}}{\partial \bar{t}} \mathrm{tr}\Big(\mathcal{F}_{\bar{I}\bar{J}}(\mathcal{Z}) \mathrm{Hol}_z(\mathcal{C}(t))\Big). \tag{5.40}$$

In Eq. 5.40, $\partial/\partial\bar{t}$ is the $\bar{\partial}$-operator on the parameter space of the holomorphic family and $\omega(\mathcal{Z})$ is the meromorphic differential on $\mathcal{C}(t)$ whose only singularities are simple poles at the intersections with residues ± 1. The notation $\mathcal{F}_{\bar{I}\bar{J}}(\mathcal{Z})$ denotes the components of the (0, 2)-form curvature,

$$\mathcal{F}_{\bar{I}\bar{J}} d\bar{\mathcal{Z}}^{\bar{I}} \wedge d\bar{\mathcal{Z}}^{\bar{J}} = \bar{\partial}\mathcal{A} + \mathcal{A} \wedge \mathcal{A}. \tag{5.41}$$

Equation (5.40) is the mathematical expression of the physical picture, familiar from Wilson loops along real curves, that any dependence on $\bar{t}$ generated by a holomorphic deformation is given by the flux of $\mathcal{F}$ through the holomorphic surface swept out by the family $\mathcal{C}(t)$.

Let us introduce a one-parameter holomorphic family of maps $\mathcal{Z}(t) : \Sigma \to \mathbb{CP}^{3|4}$. The pull-back of the partial connection to each Riemann sphere Σ_j now depends on the parameter t through the dependence on the holomorphic family $\mathcal{Z}(t)$,

$$\mathcal{A}(\sigma, t) \equiv (\mathcal{Z}_t^* \mathcal{A})(\sigma). \tag{5.42}$$

This defines a holomorphic parallel propagator on each component Σ_j by

$$(\bar{\partial}_\sigma + \mathcal{A}(\sigma, t))\mathrm{U}(\sigma, \sigma'; t) = 0 \tag{5.43}$$

and the boundary condition $\mathrm{U}(\sigma', \sigma', t) = 1$. Thus, the holomorphic parallel propagator now depends on the holomorphic family of maps $\mathcal{Z}(t)$.

Now consider the following integral

$$\int_{\Sigma_j} \omega(\sigma) \wedge \mathrm{U}(\sigma_j, \sigma; t)\, (\bar{\partial}_\sigma + \mathcal{A}(\sigma, t))\mathrm{U}(\sigma, \sigma_{j-1}; t) = 0 \tag{5.44}$$

where $\omega(\sigma)$ has only simple poles at the points σ_{j-1} and σ_j, with residues $+1$ and -1 respectively. This vanishes identically from the definition of the holomorphic frame. Now acting with $\bar{\partial}_t = \partial/\partial\bar{t}$ we find

$$\begin{aligned} 0 &= \bar{\partial}_t \left[\int_{\Sigma_j} \omega(\sigma) \wedge \mathrm{U}(\sigma_j, \sigma; t)\, \big(\bar{\partial}_\sigma + \mathcal{A}(\sigma, t)\big)\, \mathrm{U}(\sigma, \sigma_{j-1}; t) \right] \\ &= \int_{\Sigma_j} \omega(\sigma) \wedge \mathrm{U}(\sigma_j, \sigma; t)\, \big(\bar{\partial}_\sigma + \mathcal{A}(\sigma, t)\big)\, \bar{\partial}_t \mathrm{U}(\sigma, \sigma_{j-1}; t) \\ &\qquad - \int_{\Sigma_j} \omega(\sigma) \wedge \mathrm{U}(\sigma_j, \sigma; t)\, \bar{\partial}_t \mathcal{A}(\sigma, t)\, \mathrm{U}(\sigma, \sigma_{j-1}; t) \\ &= -\bar{\partial}_t \mathrm{U}(\sigma_j, \sigma_{j-1}; t) - \int_{\Sigma_j} \omega(\sigma) \wedge \mathrm{U}(\sigma_j, \sigma; t)\, \bar{\partial}_t \mathcal{A}(\sigma, t)\, \mathrm{U}(\sigma, \sigma_{j-1}; t), \end{aligned} \tag{5.45}$$

where the third line follows from integrating by parts in the first term and using the poles of $\omega(\sigma)$ to perform the integral.

In the final term of Eq. (5.45) we have

$$\bar{\partial}_t \mathcal{A}(\sigma, t) = \left(\frac{\partial \mathcal{A}_{\bar{I}}}{\partial \bar{Z}^{\bar{J}}} \frac{\partial \bar{Z}^{\bar{J}}}{\partial \bar{t}} \frac{\partial \bar{Z}^{\bar{I}}}{\partial \bar{\sigma}} + \mathcal{A}_{\bar{I}} \frac{\partial^2 \bar{Z}^{\bar{I}}}{\partial \bar{t}\, \partial \bar{\sigma}} \right) \mathrm{d}\bar{t} \wedge \mathrm{d}\bar{\sigma} . \tag{5.46}$$

The important property here is that only antiholomorphic derivatives of the partial connection arise because the family $\mathcal{Z}(t)$ depends holomorphically on the parameter t. Now integrating by parts again we find

$$\begin{aligned} \overline{D}_{\bar{t}} \mathrm{U}(\sigma_j, \sigma_{j-1}; t) &\equiv \bar{\partial}_t \mathrm{U}(\sigma_j, \sigma_{j-1}; t) + \mathcal{A}(\sigma_j, t) \mathrm{U}(\sigma_j, \sigma_{j-1}; t) - \mathrm{U}(\sigma_j, \sigma_{j-1}; t)\, \mathcal{A}(\sigma_{j-1}, t) \\ &= - \int_\Sigma \omega(\sigma) \wedge \mathrm{d}\bar{\sigma} \wedge \mathrm{d}\bar{t}\; \mathrm{U}(\sigma_j, \sigma; t)\, \mathcal{F}_{\bar{I}\bar{J}}(\sigma) \frac{\partial \bar{Z}^{\bar{I}}}{\partial \bar{\sigma}} \frac{\partial \bar{Z}^{\bar{J}}}{\partial \bar{t}} \mathrm{U}(\sigma, \sigma_{j-1}; t) \end{aligned}$$

or equivalently

$$\overline{D}_{\bar{t}}\, \mathrm{U}(\mathcal{Z}_j, \mathcal{Z}_{j-1}; t) = - \int_{X_j(t)} \omega(\mathcal{Z}) \wedge \mathrm{d}\bar{Z}^{\bar{I}} \wedge \frac{\partial Z^{\bar{J}}}{\partial \bar{t}}\; \mathrm{U}(\mathcal{Z}_j, \mathcal{Z}) \mathcal{F}_{\bar{I}\bar{J}}(\mathcal{Z})\, \mathrm{U}(\mathcal{Z}, \mathcal{Z}_{j-1}) \tag{5.47}$$

where $\overline{D}_{\bar{t}}$ is the natural covariant $\bar{\partial}$-operator on the moduli space of the holomorphic family of maps. The same conclusion can be reached by working term-by-term with the perturbative expansion of the holomorphic parallel propagator.

Equation (5.47) tells us how the holomorphic frame $\mathrm{U}(\mathcal{Z}_j, \mathcal{Z}_{j-1})$ on a single component of the momentum twistor polygon varies as we move around in the holomorphic family. For the holomorphic Wilson Loop around the whole momentum twistor polygon, we readily find

$$\begin{aligned} \frac{\partial}{\partial \bar{t}} \mathrm{W}(\mathcal{C}(t)) &= \sum_{j=1}^{n} \mathrm{Tr}\left(\left(\overline{D}_{\bar{t}} \mathrm{U}(\mathcal{Z}_{j+1}, \mathcal{Z}_j) \right) \mathrm{U}(\mathcal{Z}_j, \mathcal{Z}_{j-1}) \cdots \mathrm{U}(\mathcal{Z}_{j+2}, \mathcal{Z}_{j+1}) \right) \\ &= - \int_{\mathcal{C}(t)} \omega(\mathcal{Z}) \wedge \mathrm{d}\bar{Z}^{\bar{I}} \wedge \frac{\partial \bar{Z}^{\bar{J}}}{\partial \bar{t}}\; \mathrm{tr}_f \left(\mathcal{F}_{\bar{I}\bar{J}}(\mathcal{Z})\, \mathrm{Hol}_{\mathcal{Z}}(\mathcal{C}(t)) \right), \end{aligned} \tag{5.48}$$

which completes the proof.

5.4.2 Holomorphic Loop Equations

Equation (5.48) tells us how the holomorphic Wilson loop operator varies when we deform the curve holomorphically. For background holomorphic vector bundles $E \to \mathbb{PT}$ (classical solutions of holomorphic Chern-Simons theory) the (0,2)-form component of the curvature vanishes, $\mathcal{F}_{\bar{I}\bar{J}} = 0$, and we find

$$\frac{\partial}{\partial \bar{t}} \, \mathrm{W}(C(t)) = 0 \tag{5.49}$$

so the Wilson loop varies holomorphically.

However we are really interested in the quantum expectation value of the holomorphic Wilson loop evaluated using the full twistor action for $\mathcal{N} = 4$ super Yang-Mills theory

$$\langle \mathrm{W}(C(t)) \rangle = \int \mathrm{D}\mathcal{A} \, \mathrm{W}(C(t)) \, e^{-S(\mathcal{A})} \tag{5.50}$$

There are now corrections to Eq. (5.48) meaning that the expectation value is only meromorphic in t, developing poles when components of the curve intersect. The holomorphic *loop equations* describing this behavior may be obtained by carefully integrating by parts in the path integral.

5.4.3 Self-Dual Theory

We first consider holomorphic Chern-Simons theory, corresponding to self-dual $\mathcal{N} = 4$ super Yang-Mills theory. Acting with $\partial/\partial\bar{t}$ on the expectation value,

$$\begin{aligned} \frac{\partial}{\partial \bar{t}} \langle \mathrm{W}(C(t)) \rangle &= -\int \mathcal{D}\mathcal{A} \left[\int_{C(t)} \omega(\mathcal{Z}) \wedge \mathrm{d}\bar{\mathcal{Z}}^{\bar{I}} \wedge \frac{\partial \bar{\mathcal{Z}}^{\bar{J}}}{\partial \bar{t}} \, \mathrm{Tr}\left(\mathcal{F}_{\bar{I}\bar{J}}(\mathcal{Z}) \, \mathrm{Hol}_{\mathcal{Z}}(C(t)) \right) \right] \mathrm{e}^{-S_1[\mathcal{A}]} \\ &= \int \mathcal{D}\mathcal{A} \left[\int_{C(t)} \omega(\mathcal{Z}) \wedge \mathrm{Tr}\left(\mathrm{Hol}_{\mathcal{Z}}(C(t)) \, \frac{\delta}{\delta \mathcal{A}(z)} \, \mathrm{e}^{-S_1[\mathcal{A}]} \right) \right], \end{aligned} \tag{5.51}$$

where in the second line we have defined the variational derivative by

$$\delta S[\mathcal{A}] = \int D^{3|4}\mathcal{Z} \wedge \mathrm{Tr}\left(\delta \mathcal{A} \wedge \frac{\delta S}{\delta \mathcal{A}} \right). \tag{5.52}$$

To obtain an interesting equation for $\bar{\partial}_t \langle \mathrm{W}(C(t)) \rangle$, following [2] we integrate by parts in the path integral, bringing the variation of the connection to act on the holonomy. Essentially the same argument as above shows that under a variation of the connection at some point $\mathcal{Z}$ on the momentum twistor polygon, (displaying colour indices)

$$0=\left[\frac{\delta}{\delta\mathcal{A}(\mathcal{Z})}{}^{i}{}_{j}\left(\int_{C}\omega(\mathcal{Z}')\wedge \mathrm{U}(\mathcal{Z}_1,\mathcal{Z}')\,(\bar\partial+\mathcal{A})_C\,\mathrm{U}(\mathcal{Z}',\mathcal{Z}_0)\right)^{k}{}_{l}\right]$$
$$=-\left[\frac{\delta}{\delta\mathcal{A}(\mathcal{Z})}{}^{i}{}_{j}\,\mathrm{U}(\mathcal{Z}_1,\mathcal{Z}_0)^{k}{}_{l}\right]+\int_{C}\omega(\mathcal{Z}')\wedge \mathrm{U}(\mathcal{Z}_1,\mathcal{Z}')^{k}{}_{j}\,\bar\delta^{3|4}(\mathcal{Z},\mathcal{Z}')\,\mathrm{U}(\mathcal{Z}',\mathcal{Z}_0)^{i}{}_{l} \tag{5.53}$$

where

$$(\delta/\delta\mathcal{A}(\mathcal{Z}))^{i}{}_{j}\,\mathcal{A}(\mathcal{Z}')^{m}{}_{n}=\bar\delta^{3|4}(\mathcal{Z},\mathcal{Z}')\left(\delta^{i}{}_{n}\delta^{m}{}_{j}-\frac{1}{N}\delta^{i}{}_{j}\,\delta^{m}{}_{n}\right). \tag{5.54}$$

for SU(N) and since we are working in the planar limit the second term on the right-hand side is supressed compared to the first and has been neglected. We now apply Eq. (5.53) to the variation of the holonomy based at a point on the momentum twistor line X_1 and find

$$\mathrm{tr}\left(\frac{\delta}{\delta\mathcal{A}(\mathcal{Z})}\mathrm{Hol}_{\mathcal{Z}}(\mathcal{C}(t))\right)=\frac{\delta}{\delta\mathcal{A}(\mathcal{Z})}{}^{i}{}_{j}\left[\mathrm{U}(\mathcal{Z},\mathcal{Z}_n)^{j}{}_{k}\mathrm{U}(\mathcal{Z}_n,\mathcal{Z}_{n-1})^{k}{}_{l}\cdots\mathrm{U}(\mathcal{Z}_1,\mathcal{Z})^{m}{}_{i}\right]$$
$$=\sum_i\int_{X_i(t)}\omega(\mathcal{Z}')\wedge\bar\delta^{3|4}(\mathcal{Z},\mathcal{Z}')\,\mathrm{Tr}\Big[\mathrm{U}(\mathcal{Z},\mathcal{Z}_n)\cdots\mathrm{U}(\mathcal{Z}_{i+1},\mathcal{Z}')\Big]$$
$$\mathrm{Tr}\Big[\mathrm{U}(\mathcal{Z}',\mathcal{Z}_i)\cdots\mathrm{U}(\mathcal{Z}_1,\mathcal{Z})\Big]. \tag{5.55}$$

The factor of $\bar\delta^{3|4}(\mathcal{Z},\mathcal{Z}')$ ensures that this term has support only at points t in the moduli space where the curve $\mathcal{C}(t)$ degenerates so that the component $X_1(t)$ intersects some other component. Similar contributions arise from summing over the possible locations of the base-point $\mathcal{Z}$. In other words, $\bar\partial_t\langle\mathrm{W}(\mathcal{C}(t))\rangle$ vanishes everywhere except on boundary components of the moduli space where the holomorphic map $\mathcal{Z}(t)$ degenerates so that two points on the source curve Σ are mapped to the same image.

Combining all the terms, the loop equations for holomorphic Chern-Simons theory on twistor space (5.51) become

$$\bar\partial_t\langle\mathrm{W}(\mathcal{C}(t)])\rangle=-\int_{\mathcal{C}(t)\times\mathcal{C}(t)}\omega(\mathcal{Z})\wedge\omega(\mathcal{Z}')\wedge\bar\delta^{3|4}(\mathcal{Z},\mathcal{Z}')\,\langle\mathrm{W}(\mathcal{C}_L(t))\,\mathrm{W}(\mathcal{C}_R(t))\rangle, \tag{5.56}$$

where $\mathcal{C}_L(t)$ and $\mathcal{C}_R(t)$ are the two momentum twistor polygons obtained by ungluing $\mathcal{C}(t)$ at the new node $\mathcal{Z}=\mathcal{Z}'$ as illustrated in Fig. (5.7). Furthermore, in the planar limit, the correlation function of a product of Wilson loops factorizes into the product of correlation functions

$$\langle\mathrm{W}(\mathcal{C}_L(t))\,\mathrm{W}(\mathcal{C}_R(t))\rangle\longrightarrow\langle\mathrm{W}(\mathcal{C}_L(t))\rangle\langle\mathrm{W}(\mathcal{C}_R(t))\rangle+\mathcal{O}(1/N). \tag{5.57}$$

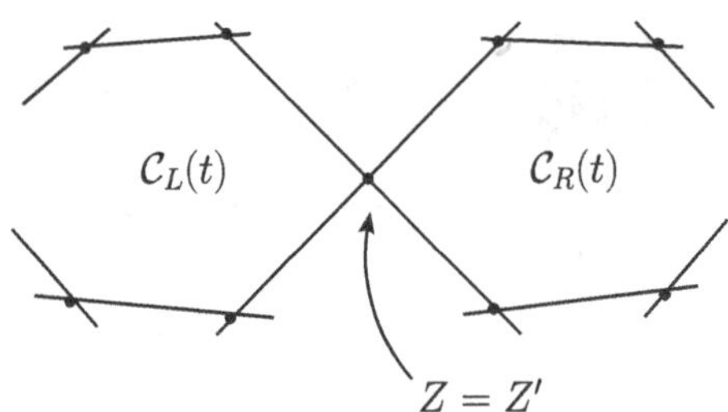

Fig. 5.7 The expectation value $\langle W(C(t)) \rangle$ in holomorphic except where the curve $C(t)$ develops self intersections

Thus, in the planar limit, the singular behaviour near the boundary of the moduli space is determined by the product of the expectation values of the Wilson Loops around the two newly created momentum twistor polygons.

An intersection of momentum twistor lines X_i and X_j corresponds precisely to the momentum space factorization channel $(p_i + \cdots + p_{j-1})^2 \to 0$. The holomorphic loop equations then imply that the holomorphic Wilson loop has the same singular behaviour as the superamplitude in this limit. In Sect. (5.5) we will make this precise and by showing the the holomorphic loop equations lead to BCFW recursion for the particular deformation.

5.4.4 Full Theory

We now consider the quantum expectation value in holomorphic Chern-Simons theory together with the interaction terms

$$S_2[\mathcal{A}] = \int d^{4|8}X \log \det(\bar{\partial} + \mathcal{A})|_X. \tag{5.58}$$

We first observe that the holomorphic Wilson loop remains a natural observable when we include the interactions. This is apparent from the expansions of the action (5.27) and of the holomorphic parallel propagator (5.26). The underlying reason for this is the close connection between determinants and holonomies. More precisely, just as

$$\delta \log \det M = \delta \operatorname{tr} \log M = \operatorname{tr}(M^{-1}\delta M) \tag{5.59}$$

for a finite dimensional matrix M, Quillen [6] showed that under a variation of the partial connection $\mathcal{A}$ on a Riemann surface Σ,

$$\delta \log \det (\bar{\partial} + \mathcal{A})\big|_\Sigma = \int_\Sigma \operatorname{tr}(J_\mathcal{A} \wedge \delta\mathcal{A}) , \tag{5.60}$$

where

$$J_\mathcal{A}(\sigma) \equiv \lim_{\sigma' \to \sigma} \left(\omega_\mathcal{A}(\sigma', \sigma) - \omega(\sigma', \sigma)\right) \tag{5.61}$$

is the limit of the difference between the Green's function for $(\bar{\partial} + \mathcal{A})$ and that for background complex structure $\bar{\partial}$. The Green's functions are non-local operators on the Riemann surface Σ; the fact that we take their limit on the diagonal can be understood as part of the trace. For a single such Green's function, the limit on the diagonal is necessarily singular, but the singularity is independent of the partial connection and cancels in the difference.

When Σ is a Riemann sphere, the Green's function may be written as

$$\omega_{\mathcal{A}}(\sigma', \sigma) = h(\sigma')\omega(\sigma', \sigma)h^{-1}(\sigma) \tag{5.62}$$

where

$$\omega(\sigma', \sigma) = \frac{\mathrm{d}\sigma}{\sigma' - \sigma}, \tag{5.63}$$

in terms of the holomorphic frame $h(\sigma)$ on Σ. Therefore Quillen's prescription reduces to

$$\begin{aligned} J_{\mathcal{A}}(\sigma) &= \lim_{\sigma' \to \sigma} \frac{\mathrm{U}(\sigma', \sigma) - \mathrm{U}(\sigma, \sigma)}{\sigma' - \sigma}\, \mathrm{d}\sigma \\ &= \left.\frac{\partial \mathrm{U}(\sigma', \sigma)}{\partial \sigma'}\right|_{\sigma' = \sigma} \mathrm{d}\sigma\,. \end{aligned} \tag{5.64}$$

The presence of the holomorphic derivative here is inconvenient. However, we see from the formal expansion (5.16) that $\mathrm{U}(\sigma', \sigma)$ depends only meromorphically on σ' and is regular as $\sigma' \to \sigma$. Thus we can rewrite the holomorphic derivative as a contour integral

$$\left.\frac{\partial \mathrm{U}(\sigma', \sigma)}{\partial \sigma'}\right|_{\sigma' = \sigma} = \oint \frac{\mathrm{d}\sigma'}{(\sigma - \sigma')^2} \mathrm{U}(\sigma', \sigma) \tag{5.65}$$

and hence

$$\begin{aligned} \delta \log \det\, (\bar{\partial} + \mathcal{A})\big|_{\Sigma} &= \int_{\Sigma} \mathrm{d}\sigma \wedge \mathrm{tr}\left(\oint \frac{\mathrm{d}\sigma'}{(\sigma - \sigma')^2} \mathrm{U}(\sigma', \sigma)\, \delta\mathcal{A}(\sigma)\right) \\ &= \int_{\Sigma \times S^1 \times S^1} \frac{\mathrm{d}\sigma \wedge \mathrm{d}\sigma' \wedge \mathrm{d}\sigma''}{(\sigma - \sigma')(\sigma' - \sigma'')(\sigma'' - \sigma)} \mathrm{tr}\left(\mathrm{U}(\sigma', \sigma)\, \delta\mathcal{A}(\sigma)\right) \end{aligned} \tag{5.66}$$

where the integrals over σ'' and σ' are performed over contours encircling the poles at $\sigma'' = \sigma'$ and $\sigma' = \sigma$, respectively. The reason for introducing the new point σ'' will become clear momentarily.

In the present context, we have a holomorphic map to twistor space $\mathcal{Z} : \Sigma \to \mathbb{CP}^{3|4}$ whose image is the line X. As in the perturbative computations, we will introduce the following parametrization of the map

$$\mathcal{Z}(\sigma) = \mathcal{Z}_a + \sigma \mathcal{Z}_b := \hat{\mathcal{Z}} \tag{5.67}$$

where

$$\mathcal{Z}(\sigma'') = \mathcal{Z}_a \qquad \mathcal{Z}(\sigma') = \mathcal{Z}_b\,. \tag{5.68}$$

In this parametrization, we have

$$\frac{\mathrm{d}\sigma \wedge \mathrm{d}\sigma' \wedge \mathrm{d}\sigma''}{(\sigma-\sigma')(\sigma'-\sigma'')(\sigma''-\sigma)} = \mathcal{Z}^*\left[\omega_{ab}(\hat{\mathcal{Z}}) \wedge \frac{\langle a\,\mathrm{d}a\rangle \wedge \langle b\,\mathrm{d}b\rangle}{\langle a\,b\rangle^2}\right] \tag{5.69}$$

where $\omega_{ab}(Z)$ is the meromorphic form with simple poles at Z_a and Z_b, and where $\langle a\,\mathrm{d}a\rangle\langle b\,\mathrm{d}b\rangle/\langle a\,b\rangle^2$ is the fundamental bi-differential on the line X with a double pole along the diagonal.

The bi-differential combines with the integral over the space of lines X, since

$$\mathrm{d}^{4|8}X \wedge \frac{\langle a\,\mathrm{d}a\rangle \wedge \langle b\,\mathrm{d}b\rangle}{\langle a\,b\rangle^2} = \mathrm{D}^{3|4}\mathcal{Z}_a \wedge \mathrm{D}^{3|4}\mathcal{Z}_b\,, \tag{5.70}$$

giving a contour integral over $\mathbb{CP}^{3|4}{}_a \times \mathbb{CP}^{3|4}{}_b$. Therefore, the variation of the new term in the action is

$$\delta \int_{\Gamma} \mathrm{d}^{4|8}X \,\log\det\,(\bar{\partial} + \mathcal{A})\big|_X$$
$$= \int_{\Gamma\times X\times S^1\times S^1} \mathrm{D}^{3|4}\mathcal{Z}_a \wedge \mathrm{D}^{3|4}\mathcal{Z}_b \wedge \omega_{ab}(\hat{\mathcal{Z}})\,\mathrm{tr}\Big(\mathrm{U}(\mathcal{Z}_b,\hat{\mathcal{Z}})\,\delta\mathcal{A}(\hat{\mathcal{Z}})\Big)\,, \tag{5.71}$$

where the $S^1 \times S^1$ component of the contour sends the twistors $\mathcal{Z}_a \to \hat{\mathcal{Z}}$ and $\mathcal{Z}_b \to \hat{\mathcal{Z}}$, differentiating $\mathrm{U}(\mathcal{Z}_b, \hat{\mathcal{Z}})$ in the process. The integral over $\mathcal{Z} \in X$ is performed using the dependence on $\hat{\mathcal{Z}}$ in the holomorphic parallel propagator and in the variation of the partial connection.

Equation (5.71) expresses the variation of the logarithm of the determinant in the twistor action in terms of a holomorphic parallel propagator on the line X. It combines beautifully with the holonomy around the curve $C(t)$, providing a new contribution to the loop equations

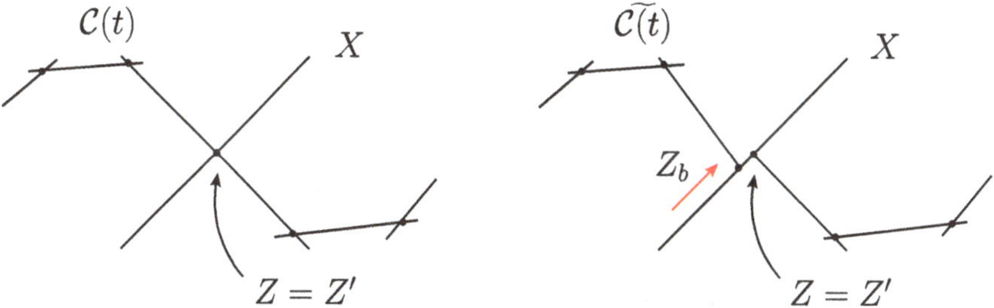

Fig. 5.8 The forward term in the loop equations can be understood as a holomorphic Wilson loop around a new curve $\widetilde{C} \cup X$, where $\widetilde{C}$ reduces to C when $\mathcal{Z}_b \to \hat{\mathcal{Z}}$ along X. In the scattering amplitude context this is interpreted as a forward limit of an $n+2$ particle amplitude

$$-\lambda \int\limits_{\Gamma \times S^1 \times S^1} \mathrm{D}^{3|4}\mathcal{Z}_a \wedge \mathrm{D}^{3|4}\mathcal{Z}_b$$

$$\left[\int\limits_{\mathcal{C}(t) \times X} \omega(\mathcal{Z}) \wedge \omega_{ab}(\hat{\mathcal{Z}}) \wedge \bar{\delta}^{3|4}(\mathcal{Z}, \hat{\mathcal{Z}})\, \mathrm{tr}\Big(\mathrm{U}(\mathcal{Z}_b, \hat{\mathcal{Z}})\, \mathrm{Hol}_{\hat{\mathcal{Z}}}(\mathcal{C}(t))\Big) \right] \quad (5.72)$$

The delta-function $\bar{\delta}^{3|4}(\mathcal{Z}, \hat{\mathcal{Z}})$ in this expression means that this term only contributes at points in the moduli space where the curve $\mathcal{C}(t)$ intersects the line X. The integral over Γ then adds up these contributions for each line X in the real Minkowski slice of chiral superspace. These integrals become the loop integrations via the lagrangian insertion procedure.

Finally, we would like to express this contribution in terms of another holomorphic Wilson loop. This is achieved by replacing the original curve by another curve $\widetilde{\mathcal{C}(t)}$ such that

$$\widetilde{\mathcal{C}(t)} \cap X = \{\hat{\mathcal{Z}}, \mathcal{Z}_b\} \in \mathbb{CP}^{3|4} \quad (5.73)$$

and such that $\widetilde{\mathcal{C}(t)} \to \mathcal{C}(t)$ as $\mathcal{Z}_b \to \hat{\mathcal{Z}}$. As a nodal curve, $\widetilde{\mathcal{C}(t)}$ has one more component than $\mathcal{C}(t)$, with one component becoming double covered in the limit. This is illustrated in Fig. (5.8).

Now, because the contour integral sends $\mathcal{Z}_b \to \hat{\mathcal{Z}}$, we can replace the holonomy around $\mathcal{C}(t)$, based at the point $\mathcal{Z} = \hat{\mathcal{Z}} = \mathcal{C}(t) \cap X$, by a product of holomorphic frames that transport us around $\widetilde{\mathcal{C}(t)}$ from $\mathcal{Z}_b$ to $\hat{\mathcal{Z}}$. The remaining factor of $\mathrm{U}(\mathcal{Z}_b, \hat{\mathcal{Z}})$ transports this frame back to $\mathcal{Z}_b$ along X. The resulting trace can be interpreted as a holomorphic Wilson loop around the new curve $\widetilde{\mathcal{C}(t)} \cup \mathrm{X}$, and hence we can make the replacement

$$\mathrm{Tr}\Big(\mathrm{U}(\mathcal{Z}_b, \hat{\mathcal{Z}})\mathrm{Hol}_{\hat{\mathcal{Z}}}(\mathcal{C}(t))\Big) \to \mathrm{W}(\widetilde{\mathcal{C}(t)} \cup X)\,, \quad (5.74)$$

inside the contour integral sending $\mathcal{Z}_b \to \hat{\mathcal{Z}}$.

Thus the complete holomorphic loop equation for the momentum twistor Wilson loop in planar $\mathcal{N}=4$ super Yang-Mills theory is

$$\begin{aligned}\bar{\partial}_t\langle \mathrm{W}(C(t))\rangle = &\int\limits_{C(t)\times C(t)} \omega(\mathcal{Z})\wedge\omega(\mathcal{Z}')\wedge\bar{\delta}^{3|4}(\mathcal{Z},\mathcal{Z}')\langle \mathrm{W}(C'(t))\rangle\langle \mathrm{W}(C''(t))\rangle \\ &+\lambda\int\limits_{\Gamma\times S^1\times S^1} \mathrm{D}^{3|4}\mathcal{Z}_a\wedge \mathrm{D}^{3|4}\mathcal{Z}_b \\ &\left[\int\limits_{C(t)\times X}\omega(\mathcal{Z})\wedge\omega_{ab}(\hat{\mathcal{Z}})\wedge\bar{\delta}^{3|4}(\mathcal{Z},\hat{\mathcal{Z}})\langle \mathrm{W}(\widetilde{C(t)}\cup X)\rangle\right]\end{aligned} \tag{5.75}$$

and, as in the original equations of Migdal & Makeenko [2], are entirely expressed in terms of quantum expectation values of holomorphic Wilson Loops. In the following section we show that for particular one-parameter families $C(t)$, the holomorphic loop equation reproduces either the all-loop BCFW recursion relation or the all-line recursion relation of section.

We would like to emphasize that the holomorphic loop equations can be immediately generalized to more general curves involving rational components of higher degree, and to multi-parameter deformations thereof. Although such generalizations might be interesting observables for further study, they do not appear to have any relationship with scattering amplitudes.

5.5 On-Shell Recursion

We now consider the one-parameter family of holomorphic curves $\mathcal{Z}(t) : \Sigma \to \mathbb{CP}^{3|4}$ obtained by deforming the image of a single node

$$\mathcal{Z}(t) : \sigma_n \longrightarrow \mathcal{Z}_n(t) = \mathcal{Z}_n + t\mathcal{Z}_{n-1}. \tag{5.76}$$

which translates the original momentum twistor $\mathcal{Z}_n$ along the line X_n as illustrated in Fig. (5.9). The complex line through the momentum twistor points $\mathcal{Z}_n(t)$ and $\mathcal{Z}_1$ will be denoted by $X_1(t)$. In the context of the kinematics of scattering amplitudes, this particular deformation is the one used to obtain the BCFW recursion relations.

The line $X_1(t)$ sweeps out the plane spanned by the momentum twistors ($\mathcal{Z}_{n-1}$, $\mathcal{Z}_n$, $\mathcal{Z}_1$), while all of the other components of $C(t)$ remain fixed. Lines and planes necessarily intersect in bosonic $\mathbb{CP}^3$ so for each $j = 3, \ldots, n-1$ there exists some value of the complex parameter t_j for which $X_1(t_j)$ intersects the component X_j. Lines and planes do not necessarily intersect in the superspace $\mathbb{CP}^{3|4}$, because they

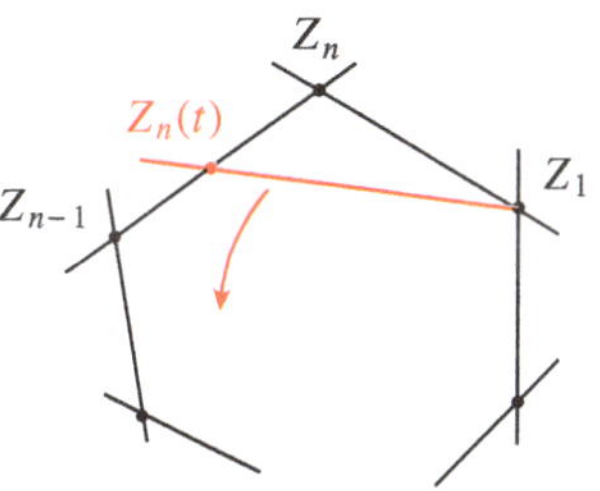

Fig. 5.9 The BCFW deformation of the momentum twistor polygon

might 'miss' in the fermionic directions. However, the fermionic δ-functions in the $\bar{\delta}^{3|4}(\mathcal{Z}, \mathcal{Z}')$ ensure that the right hand side has support only when the curves do intersect in the superspace. Let us call this intersection point $\mathcal{Z}_{I_j}$ and write $\mathcal{Z}_{n_j}$ for the point $\mathcal{Z}_n(t_j)$. Clearly, we have

$$\mathcal{Z}_{I_j} = \langle n-1, n, 1, [j-1\rangle \mathcal{Z}_j] \quad \text{and} \quad \mathcal{Z}_{n_j} = \langle j-1, j, 1, [\mathcal{Z}_{n-1}\rangle \mathcal{Z}_n] \tag{5.77}$$

by definition.

We now study the loop equations for this holomorphic family. Let us first consider the case of pure holomorphic Chern-Simons theory. Integrating over the moduli space of our holomorphic family with measure $\mathrm{d}t/t$ we find

$$-\int_{\mathbb{CP}^1} \frac{\mathrm{d}t}{t} \wedge \bar{\partial}_t \langle \mathrm{W}[C(t)] \rangle = \sum_{j=3}^{n-1} \int_{\mathbb{CP}^1 \times X_1(t) \times X_j} \frac{\mathrm{d}t}{t} \wedge \omega(\mathcal{Z}) \wedge \omega(\mathcal{Z}') \wedge \bar{\delta}^{3|4}(\mathcal{Z}, \mathcal{Z}') \langle \mathrm{W}[C'(t)] \rangle \langle \mathrm{W}[C''(t)] \rangle \tag{5.78}$$

where

$$\begin{aligned} C'(t) &= (\mathcal{Z}_{I_j}, \mathcal{Z}_1) \cup (\mathcal{Z}_1, \mathcal{Z}_2) \cup \cdots \cup (\mathcal{Z}_{j-1}, \mathcal{Z}_{I_j}) \\ C''(t) &= (\mathcal{Z}_{I_j}, \mathcal{Z}_j) \cup (\mathcal{Z}_j, \mathcal{Z}_{j+1}) \cup \cdots \cup (\mathcal{Z}_{n_j}, \mathcal{Z}_{I_j}) \end{aligned} \tag{5.79}$$

are the new momentum twistor polygons formed by the intersection. We can compute the left hand side immediately: integrating by parts gives

$$\begin{aligned} -\int_{\mathbb{CP}^1} \frac{\mathrm{d}t}{t} \wedge \bar{\partial}_t \langle \mathrm{W}(C(t)) \rangle &= \langle \mathrm{W}(C(0)) \rangle - \langle \mathrm{W}(C(\infty)) \rangle \\ &= \langle \mathrm{W}(\mathcal{Z}_1, \mathcal{Z}_2, \ldots, \mathcal{Z}_n) \rangle - \langle \mathrm{W}(\mathcal{Z}_1, \mathcal{Z}_2, \ldots, \mathcal{Z}_{n-1}) \rangle , \end{aligned} \tag{5.80}$$

the difference of holomorphic Wilson loops around the original curve and the curve where $\mathcal{Z}_n(t) \to \mathcal{Z}_{n-1}$. From a scattering amplitude perspective, this term is an inverse soft factor. Notice that the holomorphic parallel propagator $\mathrm{U}(\mathcal{Z}_n(t), \mathcal{Z}_{n-1}) \to 1$ as $\mathcal{Z}(t)_n \to \mathcal{Z}_{n-1}$ so that while the line X_n does not simply disappear from the picture, neither does it contribute to the holonomy.

On the right hand side of the equation, the dt/t moduli space integral and both integrals over the momentum twistor polygon $C(t)$ are completely fixed by the holomorphic delta-function $\bar{\delta}^{3|4}(\mathcal{Z}, \mathcal{Z}')$. These integrals can be parametrized by

$$\mathcal{Z} = \mathcal{Z}_n(t) + \sigma \mathcal{Z}_1 \quad \text{and} \quad \mathcal{Z}' = \mathcal{Z}_{j-1} + \sigma' \mathcal{Z}_j \tag{5.81}$$

in terms of local coordinates σ and σ' on $X_1(t)$ and X_j, respectively. In these coordinates, the meromorphic forms become simply

$$\omega(Z) = \frac{\mathrm{d}\sigma}{\sigma} \quad \text{and} \quad \omega(Z') = \frac{\mathrm{d}\sigma'}{\sigma'} \tag{5.82}$$

which indeed have simple poles at the nodes.

Performing the integral is now a matter of using the explicit form of the delta-function $\bar{\delta}^{3|4}(\mathcal{Z}, \mathcal{Z}')$ and computing the Jacobian. The result of the integral is simply the dual superconformal bracket

$$\int \frac{d\sigma}{\sigma}\frac{d\sigma'}{\sigma'}\frac{dt}{t}\bar{\delta}^{3|4}(\mathcal{Z}_n - t\mathcal{Z}_{n-1} + \sigma\mathcal{Z}_1, \mathcal{Z}_{j-1} + \sigma'\mathcal{Z}_j) \;=\; [n-1, n, 1, j-1, j] \tag{5.83}$$

Now including the product of the two holomorphic Wilson loops on the new curves $C'(t_j)$ and $C''(t_j)$ we find that the loop equations in pure holomorphic Chern-Simons theory become

$$\begin{aligned}\big\langle \mathrm{W}(1,\ldots,n)\big\rangle = &\big\langle \mathrm{W}(1,\ldots,n-1)\big\rangle \\ &+ \sum_{j=3}^{n-1} [n-1, n, 1, j-1, j] \\ &\big\langle \mathrm{W}(1,\ldots,j-1,\mathcal{Z}_{I_j})\big\rangle \big\langle \mathrm{W}(\mathcal{Z}_{I_j}, j, \ldots, n-1, \mathcal{Z}_{n_j})\big\rangle, \end{aligned} \tag{5.84}$$

where $\mathcal{Z}_{I_j}$ and $\mathcal{Z}_{n_j}$ were given in (5.77). This is precisely the tree-level BCFW recursion relation in momentum twistor form given in [7], but now obeyed by the holomorphic momentum twistor Wilson loop, independent of any argument involving scattering amplitudes. In other words we are seeing the factorization properties of scattering amplitudes emerging from the behaviour of the momentum twistor Wilson loop.

For the complete planar $\mathcal{N} = 4$ theory, from the holomorphic loop Eq. (5.75) we have an additional contribution

$$\lambda \int_{\Gamma\times S^1\times S^1} \mathrm{D}^{3|4}\mathcal{Z}_a \wedge \mathrm{D}^{3|4}\mathcal{Z}_b \left[\int_{\mathbb{CP}^1\times C(t)\times X} \frac{\mathrm{d}t}{t} \wedge \omega(\mathcal{Z}) \wedge \omega_{ab}(\hat{\mathcal{Z}}) \wedge \bar{\delta}^{3|4}(\mathcal{Z},\hat{\mathcal{Z}}) \big\langle \mathrm{W}(\widetilde{C(t)} \cup X)\big\rangle\right], \tag{5.85}$$

where

$$\widetilde{C(t)}\cup X = (\mathcal{Z}_1,\mathcal{Z}_2)\cup\cdots(\mathcal{Z}_{n-1},\mathcal{Z}_n(t))\cup(\mathcal{Z}_n(t),\hat{\mathcal{Z}})\cup(\hat{\mathcal{Z}},\mathcal{Z}_b)\cup(\mathcal{Z}_b,\mathcal{Z}_1) \tag{5.86}$$

is the new $(n+2)$ component curve. The integrals inside the square brackets are again completely fixed by the delta-function $\bar{\delta}^{3|4}(\mathcal{Z},\hat{\mathcal{Z}})$. The intersection point $\hat{\mathcal{Z}}$ and the point $\mathcal{Z}_n(t)$ are fixed to be

$$\hat{\mathcal{Z}}_a := \big\langle n-1, n, 1, [\mathcal{Z}_a\big\rangle \mathcal{Z}_b] \quad \text{and} \quad \mathcal{Z}_{n_{ab}} := \big\langle a, b, 1, [\mathcal{Z}_{n-1}\big\rangle \mathcal{Z}_n]. \tag{5.87}$$

and the Jacobian from the t, $\mathcal{Z}$ and $\hat{\mathcal{Z}}$ integrals again produces a dual superconformal bracket $[n-1, n, 1, a, b]$. The complete planar $\mathcal{N}=4$ loop equations thus reduce to the following

$$\begin{aligned}\big\langle \mathrm{W}(1,\ldots,n)\big\rangle &= \big\langle \mathrm{W}(1,\ldots,n-1)\big\rangle \\ &\quad + \sum_{j=3}^{n-1} [n-1, n, 1, j-1, j] \\ &\quad \big\langle \mathrm{W}(1,\ldots,j-1,\mathcal{Z}_{I_j})\big\rangle \big\langle \mathrm{W}(\mathcal{Z}_{I_j}, j, \ldots, n-1, \mathcal{Z}_{n_j})\big\rangle \\ &\quad + \lambda \int_{\Gamma\times S^1\times S^1} \mathrm{D}^{3|4}\mathrm{Z}_a \wedge \mathrm{D}^{3|4}\mathrm{Z}_b\, [n-1, n, 1, a, b] \\ &\quad \big\langle \mathrm{W}(1,\ldots,n-1,\mathcal{Z}_{n_{ab}},\hat{\mathcal{Z}}_a,\mathcal{Z}_b)\big\rangle,\end{aligned} \tag{5.88}$$

where we remind the reader that the $S^1\times S^1$ contour is taken to fix $\mathcal{Z}_a, \mathcal{Z}_b \to \hat{\mathcal{Z}}$ along X (the dual superconformal bracket has a simple zero in this limit, cancelling one of the factors of $\langle a\, b\rangle$ in the denominator of the measure).

Equation (5.88) is precisely the extension of the BCFW recursion for the all-loop integrand found in [7]. Here, we have derived it independently for the momentum twistor Wilson loop. Since scattering amplitudes are precisely determined by the same recursion relation, we have thus proved the duality.

References

1. L. Mason, D. Skinner, The complete planar S-matrix of N = 4 SYM as a Wilson loop in twistor space. JHEP **1012**, 018 (2010). http://arxiv.org/abs/1009.2225
2. Y. Makeenko, A.A. Migdal, Exact equation for the loop average in multicolor QCD. Phys. Lett. **B88**, 135 (1979)
3. M. Bullimore, D. Skinner, Holomorphic linking, loop equations and scattering amplitudes in twistor space. http://arxiv.org/abs/1101.1329
4. A. Belitsky, G. Korchemsky, E. Sokatchev, Are scattering amplitudes dual to super Wilson loops? Nucl. Phys. **B855**, 333–360 (2012). http://arxiv.org/abs/1103.3008
5. A. Belitsky, S. Caron-Huot, Superpropagator and superconformal invariants. http://arxiv.org/abs/1209.0224
6. D. Quillen, Determinants of cauchy-riemann operators over riemann surfaces. Func. Anal. Appl. **19**, 37–41 (1985)
7. N. Arkani-Hamed, J.L. Bourjaily, F. Cachazo, S. Caron-Huot, J. Trnka, The all-loop integrand for scattering amplitudes in planar N = 4 SYM. JHEP **1101**, 041 (2011). http://arxiv.org/abs/1008.2958

Chapter 6
Anomalies

Tree-level superamplitudes and the integrands of loop corrections are invariant under the Yangian of the superconformal symmetry $\mathcal{Y}(\mathfrak{psu}(2,2|4)$, which is represented in momentum twistor space by the generators [1, 2]

$$\begin{aligned} J^I{}_J &= \sum_i \mathcal{Z}_i^I \frac{\partial}{\partial \mathcal{Z}_i^J} \\ J^{(1)I}{}_J &= \sum_{i<j} (-1)^K \left[\mathcal{Z}_i^I \frac{\partial}{\partial \mathcal{Z}_i^K} \mathcal{Z}_j^K \frac{\partial}{\partial \mathcal{Z}_j^J} - (i \leftrightarrow j) \right]. \end{aligned} \tag{6.1}$$

For the integrated loop amplitudes, the situation is very much more interesting. Many symmetry generators do not annihilate loop corrections to superamplitudes. The origin of this breaking depends on the generator; some generators are broken because of infrared divergences, while some supersymmetries are broken even for IR finite observables. In all cases, the anomalies can be understood through the amplitude/Wilson loop correspondence and its supersymmetric extension, and provide powerful constraints on the amplitudes. In principle, the symmetries and their anomalies can completely determine the S-matrix.

Loop corrections to superamplitudes are infrared divergent and require regularization. For example, in dimensional regularization the loop amplitudes then depend on the dimensional regulator $d = 4 - 2\epsilon$ and an associated renormalization scale μ. This procedure breaks the dual conformal symmetries,

$$D = \sum_{i=1}^{n} \frac{1}{2} \left(\lambda_{i\alpha} \frac{\partial}{\partial \lambda_{i\alpha}} - \mu_i^{\dot{\alpha}} \frac{\partial}{\partial \mu^{\dot{\alpha}}} \right) \qquad K^{\alpha\dot{\alpha}} = \sum_{i=1}^{n} \mu^{\dot{\alpha}} \frac{\partial}{\partial \lambda_{i\alpha}}. \tag{6.2}$$

From the Wilson loop perspective, the breaking of dual conformal symmetry is a consequence of ultraviolet divergences arising at the cusps of the null polygon. Furthermore, the dual conformal generators are symmetries of the dual superspace lagrangian and their consequences can be analysed using Ward identities.

M. R. Bullimore, *Scattering Amplitudes and Wilson Loops in Twistor Space*, Springer Theses, DOI: 10.1007/978-3-319-00909-4_6,

The infrared divergent structure of planar colour-ordered amplitudes have a universal factorized and exponentiated form [3–15]. The same structure has been independently derived through the amplitude/Wilson loop correspondence in [16, 17] using the known exponentiation of ultraviolet cusp divergences of null Wilson loops [18, 19]. The infrared structure in dimensional regularization is

$$\log \mathcal{A}_n(a) = \sum_{l=0}^{\infty} a^l \left[\frac{\Gamma^{(l)}}{(l\epsilon)^2} + \frac{\Gamma^{(l)}_{\text{col}}}{l\epsilon} \right] \sum_{i=1}^{n} \left(\frac{\mu^2}{-x^2_{i,i+2}} \right)^{l\epsilon} + \mathcal{F}_n(a) + \mathcal{O}(\epsilon) , \tag{6.3}$$

where $\mathcal{F}_n(a)$ is infrared finite and independent of the renormalization scale μ and the leading infrared divergence is controlled by the cusp anomalous dimension [20, 21],

$$\Gamma(a) = \sum_{l=1}^{\infty} a^l \, \Gamma^{(l)} . \tag{6.4}$$

It should be emphasized that the derivation of this infrared divergence structure from the Wilson loop perspective has only been performed for MHV amplitudes, although it is expected to be universal. A concrete derivation from the supersymmetric extension of the Wilson loop is, in the author's opinion, an important open question.

Given the above infrared divergence structure, further constraints on the function $\mathcal{F}_n(a)$ can determined from the dual conformal Ward identities [17],

$$\begin{aligned} D \log \mathcal{F}_n(a) &= 0 \\ K^{\alpha\dot\alpha} \log \mathcal{F}_n(a) &= \frac{1}{2} \Gamma(a) \sum_{i=1}^{n} (2x_i^{\alpha\dot\alpha} - x_{i-1}^{\alpha\dot\alpha} - x_{i+1}^{\alpha\dot\alpha}) \log (x^2_{i-1,i+1}) . \end{aligned} \tag{6.5}$$

The general solution is

$$\mathcal{F}_n(a) = \Gamma(a) \mathcal{F}_n^{\text{BDS}} + \log \mathcal{R}_n(a) \tag{6.6}$$

where the particular solution $\mathcal{F}_n^{\text{BDS}}$ coincides with the BDS ansatz [22] and can be obtained from the one-loop result. The remainder function $\mathcal{R}_n(a)$ is annihilated by the dual conformal generators and consequently it depends only on the dual conformal cross-ratios formed from the cusps. It cannot be constrained further by dual conformal symmetry alone. This definition of the remainder function differs from another one commonly used in the literature, see for example [23], but appears to be more natural in the context of anomaly equations.

The remainder function $\mathcal{R}_n(a)$ is finite and dual conformally invariant. Nevertheless, some dual supersymmetries that annihilate the tree-level superamplitudes do not annihilate the remainder function [24, 25], in particular,

$$\bar{Q}_{\alpha a} \mathcal{R}_n \neq 0 \qquad S^{\dot\alpha}{}_a \mathcal{R}_n \neq 0 . \tag{6.7}$$

The aim of this chapter is to understand the reason for this anomaly and derive an anomalous Ward identity from the supersymmetric Wilson loop. Remarkably, the anomalous contribution to the Ward identity can be obtained by integrating out an additional cusp from an $(n+1)$-point Wilson loop. The resulting anomaly equations, together with anomaly equations for the level-one Yangian generators $Q^{(1)}$ and $\bar{S}^{(1)}$ and known behaviour under collinear limits, can in principle be used to determine the complete perturbative S-matrix [26].

The work in the chapter is based on Ref. [27].

6.1 Supersymmetries of Self-Dual $\mathcal{N}=4$ Yang–Mills

The dual superconformal transformations of the self-dual self-dual theory can be found very straightforwardly from the twistor formulation. The holomorphic Chern-Simons action,

$$\int \mathrm{D}^{3|4}Z \wedge \mathrm{Tr}\left(\mathcal{A}\wedge\bar{\partial}\mathcal{A}+\frac{2}{3}\mathcal{A}\wedge\mathcal{A}\wedge\mathcal{A}\right), \tag{6.8}$$

is manifestly invariant under the geometric action of the superconformal group

$$\delta\,\mathcal{A}=\epsilon^{I}{}_{J}\mathcal{Z}^{I}\frac{\partial\mathcal{A}}{\partial\mathcal{Z}^{J}}\,. \tag{6.9}$$

The dual superconformal transformations of the component fields can then be read from the component form of the non-linear Penrose transform (2.65).

Here we concentrate on the dual supersymmetry transformations, generated by

$$\begin{aligned} Q_{\alpha a} &= \lambda_\alpha\frac{\partial}{\partial\chi^a} & \bar{S}^{\dot\alpha}{}_a &= \mu^{\dot\alpha}\frac{\partial}{\partial\chi^a}\\ \bar{Q}_{\dot\alpha}{}^a &= \chi^a\frac{\partial}{\partial\mu^{\dot\alpha}} & S^{\alpha a} &= \chi^a\frac{\partial}{\partial\lambda_\alpha} \end{aligned} \tag{6.10}$$

For the Poincare supersymmetry generators, we find the following transformations of the component fields

$$\begin{aligned} \delta A &= -i\,|\epsilon^a\rangle[\bar{\psi}_a|\\ \delta|\bar{\psi}_a] &= D\phi_{ab}|\epsilon^b\rangle + i\,[\bar{\epsilon}_a|F^+\\ \delta\phi_{ab} &= -i\,\varepsilon_{abcd}\langle\epsilon^c\psi^d\rangle + i\,([\bar{\epsilon}_a\bar{\psi}_b]-[\bar{\epsilon}_b\bar{\psi}_a])\\ \delta|\psi^a\rangle &= i\,G|\epsilon^a\rangle + \frac{i}{2}\left[\phi^{ab},\phi_{bc}\right]|\epsilon^c\rangle + [\bar{\epsilon}_b|D\phi^{ab}\\ \delta G_{\alpha\beta} &= \epsilon^a_{(\alpha}\left[\psi^b_{\beta)},\phi_{ab}\right] + [\bar{\epsilon}_a|D_{(\alpha}\psi^a_{\beta)}\quad, \end{aligned} \tag{6.11}$$

where the parameters $|\epsilon^a\rangle$ and $|\bar{\epsilon}_a]$ correspond to the supersymmetries $|Q_a\rangle$ and $|\bar{Q}^a]$. There are similar expressions for the conformal supersymmetry generators $|S^a\rangle$ and $|\bar{S}_a]$. It is straightforward to check that the transformations (6.11) leave invariant the space-time action for the self-dual theory

$$S_1 = \int \mathrm{d}^4x \,\mathrm{Tr}\left(-\frac{1}{4}G^{\alpha\beta}F_{\alpha\beta} + i\bar{\psi}_{\dot{\alpha}a}D^{\dot{\alpha}\alpha}\psi^a_\alpha + \frac{1}{4}D_{\alpha\dot{\alpha}}\phi^{ab}D^{\alpha\dot{\alpha}}\phi_{ab} + \bar{\psi}_{\dot{\alpha}a}[\phi^{ab}, \bar{\psi}^{\dot{\alpha}}_b]\right) \tag{6.12}$$

up to boundary terms, and form a representation of the supersymmetry algebra up to self-dual equations of motion and field dependent gauge transformations.

6.2 Supersymmetries of Full $\mathcal{N} = 4$ Yang–Mills

Now we consider the supersymmetry transformations of the complete $\mathcal{N} = 4$ super Yang-Mills theory. We demonstrate that the self-dual supersymmetry generators $|\bar{Q}^a\rangle$ and $|S^a\rangle$ do not leave invariant the additional interaction terms in the lagrangian. These supersymmetry generators alone receive a correction at order g^2. This observation is the fundamental reason why these dual supersymmetry generators are broken in loop corrections to scattering amplitudes, even for the infra-red finite remainder function.

The non-self-dual correction to the generators would look highly non-local from the twistor space perspective. Hence we will find the correction by checking invariance of the full space-time action. In the introduction, we explained that full $\mathcal{N} = 4$ super Yang-Mills has an expansion around the self-dual sector

$$S = S_1 + \mathrm{g}^2\, S_2 \tag{6.13}$$

where the self-dual action was given in Eq. (6.12) and the non-self-dual interaction term is

$$S_2 = \int \mathrm{d}^4x \;\mathrm{Tr}\left(-\frac{1}{2}G^{\alpha\beta}G_{\alpha\beta} + \psi^{\alpha a}[\phi_{ab}, \psi_\alpha{}^b] + \frac{1}{8}[\phi^{ab}, \phi^{cd}][\phi_{ab}, \phi_{cd}]\right) . \tag{6.14}$$

The non-self-dual interaction term depends only on the component fields $\{\phi, \psi, G\}$ in one-half of the supermultiplet. This observation will be important momentarily.

However, let us first make two elementary but important observations. This first is that by integrating out the auxiliary field using the equation of motion $F_{\alpha\beta} = \mathrm{g}^2 G_{\alpha\beta}$, we obtain

$$\int d^4x \;\mathrm{Tr}\left(-\frac{1}{4\mathrm{g}^2}F_{\alpha\beta}F^{\alpha\beta}\right) , \tag{6.15}$$

which is the standard gauge field kinetic term up to a total derivative. The second is that the $|Q_a\rangle$ supersymmetry transformation of the auxiliary field agrees with the standard transformation of the self-dual field strength $F_{\alpha\beta}$ upon applying the equation of motion for $\psi_\alpha{}^a$,

$$\delta F_{\alpha\beta} = \mathrm{i}\epsilon_{(\alpha}{}^a D_{\beta)}{}^{\dot{\alpha}}\,\bar{\psi}_{\dot{\alpha}a}\,. \tag{6.16}$$

It is important, however, that the $|Q_a\rangle$ supersymmetry transformations in Eq. (6.11) leave invariant the full action without imposing any of the equations of motion.

Let us now consider how the supersymmetry generators $|\bar{Q}^a]$ and $|S^a\rangle$ are corrected in the full theory. For our purposes, we need only the correction Poincare supersymmetry generator $|\bar{Q}^a]$, since the corresponding results for $|\bar{S}_a\rangle$ will follow for infra-red finite quantities by commuting with the dual special conformal generator $K^{\alpha\dot{\alpha}}$. The correction acts trivially on the fields ϕ, ψ and the auxiliary field G, but non-trivially on the gluon and the positive chirality gluino:

$$\begin{aligned} \delta^{(1)} A &= \mathrm{i}\mathrm{g}^2 |\psi^a\rangle[\bar{\epsilon}_a| \\ \delta^{(1)}|\bar{\psi}_a] &= -\frac{\mathrm{i}}{2}\mathrm{g}^2\,|\bar{\epsilon}_c]\left[\phi^{cb},\phi_{ab}\right]. \end{aligned} \tag{6.17}$$

Since the non-self-dual interaction terms S_2 depend only on the fields $\{\phi, \psi, G\}$ that are unaffected by the correction, we immediately see that $\delta^{(1)} S_2 = 0$. Invariance of the self-dual action under the self-dual supersymmetries, $\delta^{(0)} S_1 = 0$, and of the full action under the full supersymmetries then implies that

$$\delta^{(1)} S_1 + \mathrm{g}^2\,\delta^{(0)} S_2 = 0 \tag{6.18}$$

which may be readily checked. Note that only this combination vanishes. In other words, the self-dual theory is not invariant under the supersymmetries of the full theory, nor is the full theory invariant under the supersymmetries of the self-dual theory.

Note that we have chosen a normalization of the fields so that perturbation theory in g^2 can be treated as an expansion away from the self-dual, rather than free, theory. If desired, the standard normalization with $1/\mathrm{g}^2$ in front of the whole action can be obtained by rescaling

$$A \to A \quad |\bar{\psi}_a] \to \mathrm{g}^{-\frac{1}{2}}|\bar{\psi}_a] \quad \phi_{ab} \to \mathrm{g}^{-1}\,\phi_{ab} \quad \langle\psi^a| \to \mathrm{g}^{-\frac{3}{2}}\langle\psi^a| \quad G \to \mathrm{g}^{-2}G\,. \tag{6.19}$$

The further rescaling

$$|\theta^a\rangle \to \mathrm{g}^{-\frac{1}{2}}|\theta^a\rangle \qquad |\bar{\theta}_a] \to \mathrm{g}^{\frac{1}{2}}|\bar{\theta}_a] \qquad \langle\epsilon^a| \to \mathrm{g}^{-\frac{1}{2}}\langle\epsilon^a| \qquad |\bar{\epsilon}_a] \to \mathrm{g}^{\frac{1}{2}}|\bar{\epsilon}_a] \tag{6.20}$$

means that the supersymmetry transformations become independent of the coupling.

6.3 A Ward Identity for Super Wilson Loops

In this section, we review the space-time version of the superamplitude/super Wilson loop correspondence in planar $\mathcal{N} = 4$ supersymmetric Yang-Mills theory. When evaluated in self-dual planar theory, the supersymmetric Wilson loop corresponds to the tree-level superamplitude, while the same correlator taken in full planar $\mathcal{N} = 4$ SYM is dual to the full quantum superamplitude. Thus, with the action normalized as $S = S_1 + \mathrm{g}^2 S_2$, the perturbative expansion of the superloop correlator matches that of the amplitude order by order in g^2. The source of the anomaly is then precisely the difference between the self-dual and full supersymmetry generators.

6.3.1 The Superconnection

Our starting point is the null polygonal superloop correlator

$$\mathrm{W}[C_n] \equiv \frac{1}{N}\left\langle \mathrm{Tr}\,\mathrm{P}\exp\left(\mathrm{i}\oint_{C_n} \mathbb{A}\right)\right\rangle \tag{6.21}$$

which is the expectation value of the trace of the holonomy in the fundamental representation of a superconnection

$$\mathbb{A}(x,\theta) = \mathcal{A}_{\alpha\dot{\alpha}}(x,\theta)\mathrm{d}x^{\alpha\dot{\alpha}} + \mathcal{A}_{\alpha a}(x,\theta)\mathrm{d}\theta^{\alpha a} \tag{6.22}$$

on chiral superspace. This superconnection is determined by the constraints

$$\begin{aligned} \lambda^\alpha\lambda^\beta\left[\nabla_{\dot{\alpha}\alpha}\,,\nabla_{\beta b}\right] &= 0 \\ \lambda^\alpha\lambda^\beta\left\{\nabla_{a\alpha}\,,\nabla_{\beta b}\right\} &= 0 \end{aligned} \tag{6.23}$$

that state that $\mathbb{A}$ is integrable along α-planes in chiral superspace, together with a supergauge fixing condition $\theta^{\alpha a}\mathcal{A}_{\alpha a} = 0$. The supercovariant derivatives with respect to the superconnection are defined by $\nabla_{\alpha\dot{\alpha}} = \partial_{\alpha\dot{\alpha}} - \mathrm{i}\mathcal{A}_{\alpha\dot{\alpha}}$ and $\nabla_{\alpha a} = \partial/\partial\theta^{\alpha a} - \mathrm{i}\mathcal{A}_{\alpha a}$, whereas the regular covariant derivative will always be denoted by $D_{\alpha\dot{\alpha}} = \partial_{\alpha\dot{\alpha}} - \mathrm{i}A_{\alpha\dot{\alpha}}$. The component expansions of the superconnection up to fourth order can be found in the appendix.

6.3.2 The Supercurvatures

In space-time, the integrability conditions mean that the non-vanishing components of the supercurvature are

$$
\begin{aligned}
\left[\nabla_{\alpha\dot{\alpha}}, \nabla_{\beta\dot{\beta}}\right] &= i\,\varepsilon_{\alpha\beta}\mathcal{F}^{+}_{\dot{\alpha}\dot{\beta}} + i\,\varepsilon_{\dot{\alpha}\dot{\beta}}\mathcal{F}^{-}_{\alpha\beta} \\
\left[\nabla_{\alpha\dot{\alpha}}, \nabla_{\beta b}\right] &= i\,\varepsilon_{\alpha\beta}\mathcal{F}_{\dot{\alpha}b} \\
\left\{\nabla_{\alpha a}, \nabla_{\beta b}\right\} &= i\,\varepsilon_{\alpha\beta}\mathcal{F}_{ab}
\end{aligned}
\tag{6.24}
$$

where in the first line, we have decomposed the bosonic supercurvature into its self-dual and anti self-dual components. The component expansion of the supercurvatures maybe calculated from that of the superconnection and are presented in the appendix. Here we note only that the lowest components of these supercurvatures are the corresponding fundamental component fields with the same quantum numbers,

$$
\mathcal{F}_{\dot{\alpha}\dot{\beta}} = F_{\dot{\alpha}\dot{\beta}} + \cdots \qquad \mathcal{F}_{\dot{\alpha}a} = \bar{\psi}_{\dot{\alpha}a} + \cdots \qquad \mathcal{F}_{ab} = \phi_{ab} + \cdots\,. \tag{6.25}
$$

By taking further covariant derivatives of the supercurvature, we obtain superfields

$$
\begin{aligned}
\mathcal{F}_{\alpha}{}^{a} &:= \frac{1}{3!}\epsilon^{abcd}\nabla_{\alpha b}\mathcal{F}_{cd} \\
\mathcal{G}_{\alpha\beta} &\equiv \frac{1}{4!}\epsilon^{abcd}\nabla_{\alpha a}\nabla_{\beta b}\mathcal{W}_{cd}
\end{aligned}
\tag{6.26}
$$

whose lowest components are the remaining component fields $\psi_{\alpha}{}^{a}$ and $G_{\alpha\beta}$, respectively. We also note that each component of the self-dual field $\mathcal{F}_{\alpha\beta}$ is proportional to an equation of motion of the self-dual action.

It is straightforward to verify from the expansions that the supercurvatures obey the same equations of motion (2.66) as the corresponding component fields with covariant derivatives replaced by super-covariant derivatives. Particularly relevant for below are

$$
\begin{aligned}
\mathcal{F}^{-}_{\alpha\beta} &= g^{2}\,\mathcal{G}_{\alpha\beta} \\
\nabla^{\alpha\dot{\alpha}}\mathcal{F}_{\dot{\alpha}a} &= g^{2}\,\mathrm{i}\,[\mathcal{F}_{ab}, \mathcal{F}^{\alpha a}]\,.
\end{aligned}
\tag{6.27}
$$

The first equation in particular actually contains all of the component equations of motion; terms in the component expansion of $\mathcal{F}_{\alpha\beta}$ are proportional to self-dual equations of motion, and terms in $\mathcal{G}_{\alpha\beta}$ are proportional to the corresponding non-self-dual corrections and involve only the fields $\{\phi, \psi, G\}$ in a chiral half of the supermultiplet.

6.3.3 $\bar{Q}$ Transformation of Superconnection

Now consider how the superconnection transforms under the self-dual $\bar{Q}$ transformation of the component fields in equations. The variation is tangent to the affine space of superconnections, and hence should be constructed solely from the gauge covariant supercurvatures. From the component expansions, at least up to second order in the fermions, we find that

$$\begin{aligned}
\left[\bar{\varepsilon}\cdot\bar{\mathcal{Q}}^{(0)}, \mathcal{A}_{\alpha\dot{\alpha}}(x,\theta)\right] &= \theta_\alpha{}^b\,\bar{\epsilon}^{\dot{\beta}}{}_b\mathcal{F}_{\dot{\alpha}\dot{\beta}} + \nabla_{\alpha\dot{\alpha}}\,\omega + \frac{1}{2}\theta_\alpha{}^a\bar{\epsilon}_{\dot{\alpha}a}\langle\theta^b|\nabla|\mathcal{F}_b]\\
\left[\bar{\varepsilon}\cdot\bar{\mathcal{Q}}^{(0)}, \mathcal{A}_{\alpha a}(x,\theta)\right] &= \bar{\epsilon}^{\dot{\beta}}{}_a\theta_\alpha{}^b\mathcal{F}_{\dot{\beta}b} + \nabla_{\alpha a}\,\omega .
\end{aligned} \tag{6.28}$$

where

$$\omega \equiv \frac{1}{2}\langle\theta^a\theta^b\rangle[\bar{\epsilon}_a\mathcal{F}_b] + \cdots \tag{6.29}$$

is a field-dependent super gauge transformation and

$$\langle\theta^a|\nabla|\mathcal{F}_a] = \langle\theta^a|D|\bar{\psi}_b] + \langle\theta^a\theta^b\rangle\left(\Box\phi_{ab} - \left\{\bar{\psi}^{\dot{\alpha}}{}_a, \bar{\psi}_{\dot{\alpha}b}\right\}\right) + \cdots \tag{6.30}$$

is proportional to self-dual equations of motion. For the non-self-dual correction we find that the superconnection transforms as follows

$$\begin{aligned}
\left[\bar{\varepsilon}\cdot\bar{\mathcal{Q}}^{(1)}, A_{\alpha\dot{\alpha}}(x,\theta)\right] &= \mathrm{g}^2\,[\bar{\epsilon}_a|\mathrm{d}x|\big(|\mathcal{F}^a\rangle - \mathcal{G}|\theta^a\rangle + \frac{\mathrm{i}}{2}[\mathcal{F}_{bc}, \langle\theta^b\mathcal{F}^c\rangle]|\theta^a\rangle\big)\\
\left[\bar{\varepsilon}\cdot\bar{\mathcal{Q}}^{(1)}, \mathcal{A}_{\alpha a}(x,\theta)\right] &= 0
\end{aligned} \tag{6.31}$$

The transformation of the fermionic part of the superconnection is invariant because it is constructed only from the component fields $\{\phi, \psi, G\}$.

6.3.4 $\bar{Q}$ Ward Identity

We now consider the effect of a supersymmetry transformation

$$\begin{aligned}
\bar{\mathcal{Q}}_{\dot{\alpha}}{}^a &= \sum_i \chi_i^a \frac{\partial}{\partial\mu_i^{\dot{\alpha}}}\\
&= \sum_i\left(\theta_i^{\alpha a}\frac{\partial}{\partial x_i^{\alpha\dot{\alpha}}} + \eta_i^a\frac{\partial}{\partial\bar{\lambda}_i^{\dot{\alpha}}}\right)
\end{aligned} \tag{6.32}$$

on the external data of the superloop. The aim is to find a Ward identity relating the action on the external data of the superloop correlation function to the supersymmetry transformations of the fields inside the correlation function.

The external supersymmetry transformation (6.32) translates the contour through chiral superspace $\mathbb{CM}^{4|8}$, whilst preserving the null separation constraints,

$$(x_{i+1} - x_i)^2 = 0 \qquad (x_{i+1} - x_i)|\theta_{i+1} - \theta_i\rangle = 0 . \tag{6.33}$$

By standard techniques, any infinitesimal deformation of the contour of the superloop leads to an insertion of the supercurvature. Applying this to the deformation above we find that

$$\bar{Q}_{\dot{\alpha}}{}^{a}\,\mathrm{W}[C_n] = \frac{\mathrm{i}}{N}\oint \mathrm{d}x^{\beta\dot{\beta}}\,\theta^{\alpha a}\left\langle \mathrm{Tr}\,\mathrm{P}\left(\varepsilon_{\alpha\beta}\mathcal{F}_{\dot{\alpha}\dot{\beta}} + \varepsilon_{\dot{\alpha}\dot{\beta}}\mathcal{F}_{\alpha\beta}\right)\mathrm{Hol}[C_n]\right\rangle + \frac{\mathrm{i}}{N}\oint \mathrm{d}\theta^{\beta b}\,\theta^{\alpha a}\left\langle \mathrm{Tr}\,\mathrm{P}\left(\varepsilon_{\alpha\beta}\mathcal{F}_{\dot{\alpha}a}\right)\mathrm{Hol}[C_n]\right\rangle \tag{6.34}$$

where the fermionic-fermionic components of the supercurvature do not arise because the deformation acts trivially on the θ s. While the above deformation formula holds in both the self-dual and full theories, its consequences are strikingly different in each case.

Let us begin by considering the superloop correlator planar self-dual $\mathcal{N} = 4$ supersymmetric Yang-Mills theory. Here we make an assumption that there is a regulator for the superloop such that the self-dual field equations

$$\mathcal{F}_{\alpha\beta} = 0 \quad \text{and} \quad \nabla^{\alpha\dot{\alpha}}\mathcal{F}_{\dot{\alpha}a} = 0 \tag{6.35}$$

hold as operator equations and that the superloop correlator is invariant under supergauge transformations. Then, using the transformation of the superloop (6.28) and the insertion formula (6.34), we find that

$$\begin{aligned}(\bar{\varepsilon}\cdot\bar{Q})\,\mathrm{W}[C] &= \mathrm{i}\oint_C \left\langle \mathrm{Tr}\,\mathrm{P}\left(\left[\,\bar{\varepsilon}\cdot\bar{Q}^{(0)}, \mathbb{A}\right]\right)\mathrm{Hol}[C]\right\rangle \\ &= \frac{1}{N}\left\langle\left[\,\bar{\varepsilon}\cdot\bar{Q}^{(0)}, \mathrm{Tr}\,\mathrm{P}\exp\left(\mathrm{i}\oint_C \mathbb{A}\right)\right]\right\rangle \\ &= 0\end{aligned} \tag{6.36}$$

where the last equality follows since $\bar{Q}^{(0)}$ generates a symmetry of the self-dual theory. Recalling that the self-dual superloop reproduces the classical S-matrix, we have simply recovered the fact that tree amplitudes are annihilated by the dual supersymmetry transformation $\bar{Q}$.

For the superloop correlator in full planar $\mathcal{N} = 4$ supersymmetric Yang-Mills theory, the above Ward identity receives an important anomalous contribution. There are two sources to the correction. First of all, the field equations of the self-dual theory receive corrections,

$$\begin{aligned}\mathcal{F}_{\alpha\beta} &= \mathrm{g}^2\,\mathcal{G}_{\alpha\beta} \\ \nabla^{\alpha\dot{\alpha}}\mathcal{F}_{\dot{\alpha}a} &= g^2\,\mathrm{i}\,[\mathcal{F}_{ab}, \mathcal{G}^{\alpha a}]\end{aligned} \tag{6.37}$$

so that, for example, we cannot ignore the presence of the supercurvature $\mathcal{F}_{\alpha\beta}$ in the insertion formula. Secondly, we must now relate the curvature insertions to the complete $\bar{Q}$ transformations of the superconnection in the full theory. We now find

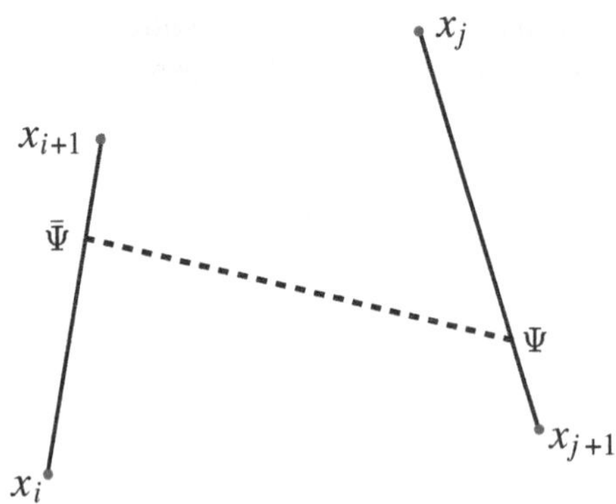

Fig. 6.1 In the Abelian theory, at order g^2, the $\bar{Q}$ anomaly receives contributions from single fermion and gauge field propagators stretched between the two copies of the framed loop

that the curvature insertions and $\bar{Q}$ transformation of the superconnection conspire together to produce an insertion of the combination $\mathcal{F}_{\alpha\beta} + g^2\mathcal{G}_{\alpha\beta}$, which vanishes now by the full equations of motion, leaving the formula

$$(\bar{\varepsilon} \cdot Q)W[C] = \frac{ig^2}{N}\left\langle \oint \mathrm{Tr}\,\mathrm{P}\left([\bar{\epsilon}_a|\mathrm{d}x|\mathcal{F}^a\rangle\, \mathrm{Hol}[C]\right)\right\rangle . \tag{6.38}$$

This involves an insertion of the curvature $\mathcal{F}_\alpha{}^a = \psi_\alpha{}^a + \cdots$ whose lowest component is the left-handed fermion. In the following sections, we will reexpress the insertion formula as an integral of an $(n + 1)$-point superloop (Fig. 6.1).

Although we have focussed on the anomaly in dual Poincaré supersymmetry generator $|\bar{Q}^a]$, a similar story is true for the dual superconformal symmetry $|S^a\rangle$. For objects invariant under dual conformal symmetry such as the remainder function $\mathcal{R}_n$, the corresponding anomaly for $|S^a\rangle$ can be determined from the superconformal algebra,

$$[K^{\alpha\dot{\beta}}\,,\; \bar{Q}_{\dot{\alpha}}^{\,a}] = \delta_{\dot{\alpha}}^{\,\dot{\beta}}\, S^{\alpha a}\,. \tag{6.39}$$

Alternatively, we can simply promote the dual supersymmetry generator to a twistor generator with components $\bar{\mathcal{Q}}^a = (S^{\alpha a},\; \bar{Q}_\alpha{}^a)$. The anomaly equation will then have an immediate extension to the twistor generator (Fig. 6.2).

6.3.5 Abelian Theory

In this subsection, we perform a simple but concrete check of the Ward identity in the abelian theory. In this case, the supercurvature insertion appearing in the anomalous contribution terminates as second order

$$\mathcal{F}_\alpha{}^a = \psi_\alpha{}^a + G_{\alpha\beta}\,\theta^\beta \tag{6.40}$$

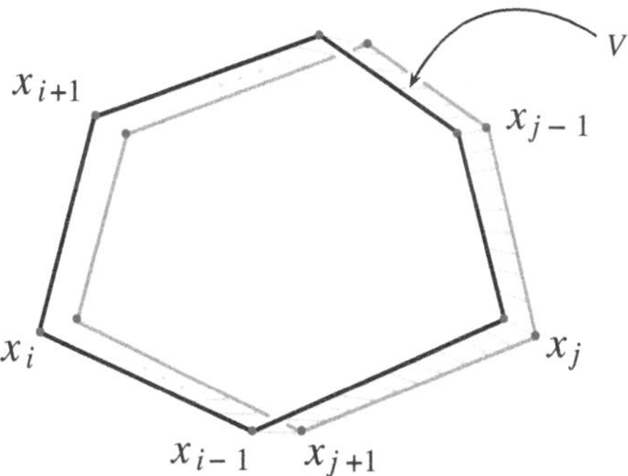

Fig. 6.2 The framed Wilson Loop. To 1-loop order, we only need consider two copies of the Wilson Loop, obtained by translating the original polygon along a nowhere null normal vector field v. In our conventions, the vertices of the original Wilson Loop are labelled by $\{x_i\}$, whereas the vertices of the translated loop are labelled by $\{x_j\}$. Since v is nowhere null, $x_{i,j}^2 \neq 0$

and hence the Ward identity for the superloop becomes

$$(\bar{\varepsilon} \cdot \bar{Q})\, \mathrm{W}[C] = \frac{\mathrm{i}g^2}{N} \left\langle \oint \left([\bar{\epsilon}_a|\mathrm{d}x|\psi^a\rangle + [\bar{\epsilon}_a|\mathrm{d}x|G\theta^a\rangle \right) \exp\left(\mathrm{i} \oint \mathbb{A} \right) \right\rangle . \tag{6.41}$$

We will now compute the right-hand side to leading order in the θ expansion, which corresponds to the $\bar{Q}$ anomaly of the one-loop MHV amplitude. To this order, the expansion of the superconnection is given

$$\mathbb{A}(x, \theta) = A + \mathrm{i}[\bar{\psi}|\mathrm{d}x|\theta^a\rangle + \mathcal{O}(\theta^2) . \tag{6.42}$$

Thus the only possible contributions are from a single propagator, connected either between the explicit fermion insertion $|\psi^a\rangle$ and the $|\bar{\psi}_a]$ from the expansion of the superloop, or between the explicit insertions of $G_{\alpha\beta}$ and a gauge field $A_{\alpha\dot{\alpha}}$ from the superloop (Fig. 6.1).

The contributions from single propagators diverge for propagators joining adjacent edges. This is expected because the one-loop amplitude is itself IR divergent. In the abelian theory and at one-loop, it is convenient to regularize by framing the superloop (Fig. 6.2). This means that we compute the cross-correlator [1]

$$\frac{\langle \mathrm{W}[C]\, \mathrm{W}[C'] \rangle}{\langle \mathrm{W}[C] \rangle \langle \mathrm{W}[C'] \rangle} \tag{6.43}$$

where the second contour is obtained by an infinitesimal deformation along a normal space-like vector field $v^{\alpha\dot{\alpha}}$. Introducing the convention that the cusps of C and C' are labelled $\{x_i\}$ and $\{x_j\}$ respectively, then we have

[1] In the non-Abelian theory, framing regularization does not simply reduce to a cross-correlator.

$$(x_i - x_j)^2 \neq 0 \qquad \forall i, j = 1, \ldots, n\,. \tag{6.44}$$

An important property of this regularization is that it preserves the Q and $\bar{S}$ supersymmetries of the chiral superloop [28].

We consider first the fermion propagator

$$\left\langle \psi_\alpha{}^a(x)\bar{\psi}_{b\dot{\beta}}(y)\right\rangle = \mathrm{i}\delta_b{}^a \frac{(x-y)_{\alpha\dot{\beta}}}{(x-y)^4}\,. \tag{6.45}$$

Performing the integrals around both copies of the loop, the Ward identity gives a lowest order contribution

$$\begin{aligned}\left\langle \int_{x_i}^{x_{i+1}} [\bar{\epsilon}_a|\mathrm{d}x|\psi^a(x)\rangle \int_{x_j}^{x_{j+1}} [\bar{\psi}_b(y)|\mathrm{d}y|\theta^b\rangle \right\rangle &= \int_0^1 \mathrm{d}s \int_0^1 \mathrm{d}t\, \frac{[i\,\bar{\epsilon}_a]\,\chi_j^a\,\langle i|x_{ij}|j]}{(x_{ij}+sx_{i+1\,i}-ty_{j+1\,j})^4} \\ &= \frac{(i-1,i,i+1,\bar{\epsilon}_a)}{(i-1,i,i+1,j)}\,\chi_j^a \log\frac{x_{i\,j+1}^2 x_{i+1\,j}^2}{x_{ij}^2\,x_{i+1\,j+1}^2}\end{aligned} \tag{6.46}$$

for each pair of edges i and j. Similarly, inserting the photon propagator

$$\left\langle G_{\alpha\beta}(x)A_{\gamma\dot{\gamma}}(y)\right\rangle = \mathrm{i}\frac{(x-y)_{\dot{\gamma}(\alpha}\varepsilon_{\beta)\gamma}}{(x-y)^4}\,, \tag{6.47}$$

one finds a contribution

$$\begin{aligned}= \left[\frac{(\bar{\epsilon}_a,i,i+1,j)}{(i-1,i,i+1,j)}\chi_{i-1}^a + \frac{(i-1,\bar{\epsilon}_a,i+1,j)}{(i-1,i,i+1,j)}\chi_i^a + \frac{(i-1,i,\bar{\epsilon}_a,j)}{(i-1,i,i+1,j)}\chi_{i+1}^a\right] \\ \log\frac{x_{i\,j+1}^2 x_{i+1\,j}^2}{x_{ij}^2\,x_{i+1\,j+1}^2}\end{aligned} \tag{6.48}$$

up to terms that cancel telescopically around the loop.

The logarithm of the cross-ratio is common to both contributions (6.46) and (6.48). The combined pre-factor is recognized as

$$\begin{aligned}\bar{\varepsilon}\cdot\bar{Q}\big(\log(i-1,i,i+1,j)\big) = \\ \left[\frac{(i-1,i,i+1,\bar{\epsilon}_a)}{(i-1,i,i+1,j)}\chi_j^a + \frac{(\bar{\epsilon}_a,i,i+1,j)}{(i-1,i,i+1,j)}\chi_{i-1}^a \right. \\ \left. + \frac{(i-1,\bar{\epsilon}_a,i+1,j)}{(i-1,i,i+1,j)}\chi_i^a + \frac{(i-1,i,\bar{\epsilon}_a,j)}{(i-1,i,i+1,j)}\chi_{i+1}^a\right].\end{aligned} \tag{6.49}$$

Thus, summing over i and j running around the two copies of the framed loop, we find that the total contribution to the anomaly is

$$(\bar{\varepsilon}\cdot\bar{Q})\mathrm{W}[C]\big|_{\mathrm{g}^2} = \mathrm{g}^2\sum_{i,j}\bar{\varepsilon}\cdot\bar{Q}\big(\log(i-1,i,i+1,j)\big)\log\frac{x_{i\,j+1}^2 x_{i+1\,j}^2}{x_{ij}^2\,x_{i+1\,j+1}^2} \tag{6.50}$$

This expression is always well-defined in four dimensions, since $x_{ij}^2 \neq 0$ for all i, j by construction. We should also include a copy of the sum with i and j interchanged to account for the fact that in framing regularization, we could perform the supersymmetry variation on either $\mathrm{W}[C]$ or $\mathrm{W}[C']$.

Now we can immediately deduce that the total derivative of the framed MHV Wilson Loop is

$$\begin{aligned}\mathrm{d}\,\mathrm{W}[C]\,|_{\mathrm{g}^2} &= \sum_{i,j}\log\frac{X_i\cdot X_{j+1}\;X_{i+1}\cdot X_j}{X_i\cdot X_j\;X_{i+1}\cdot X_{j+1}}\,\mathrm{d}\log(i-1,i,i+1,j) + (i\leftrightarrow j)\\ &= \sum_{i,j}\log(X_i\cdot X_j)\,\mathrm{d}\log\frac{(i-1,i,i+1,j-1)(i-2,i-1,i,j)}{(i-1,i,i+1,j)(i-2,i-1,i,j-1)} + (i\leftrightarrow j)\,.\end{aligned} \tag{6.51}$$

which exactly agrees with the total derivative of the one-loop cross-correlator of two bosonic Wilson loops computed in [29] (see also [30]). As the framing is removed, terms with $|i-j|>1$ remain finite, while adjacent terms diverge logarithmically reproducing the infra-red divergences of the one-loop amplitude.

6.4 Descent Equations

In the abelian theory, it was straightforward to compute the $\bar{Q}$ anomaly directly, although the calculation was no simpler than the original computations of the one-loop MHV amplitude itself, see for example [16, 31]. To understand the $\bar{Q}$ anomaly beyond one-loop, or beyond the MHV sector, we must consider the non-abelian theory. A direct computation is then considerably less appealing, both because the supercurvature insertion $\mathcal{F}_\alpha{}^a(x,\theta)$ is more complicated, and since many more diagrams contribute to the correlation function.

6.4.1 The Integral Formula

In this section, we reformulate the right hand side of the Ward identity as an integral that is carried out purely at the level of the momentum twistor data $\{\mathcal{Z}_1,\ldots,\mathcal{Z}_n\}$. Specifically, we show that the anomalous term in the Ward identity can

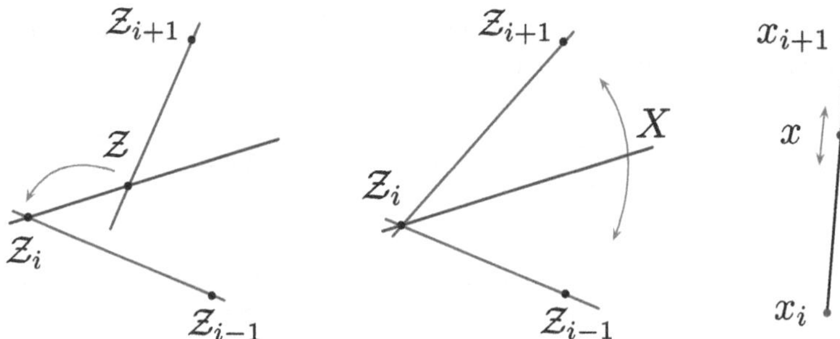

Fig. 6.3 The $(n+1)$-point superloop is integrated over a contour that fixes $\mathcal{Z} \to \mathcal{Z}_i$ and also causes the line X to move in the plane $(i-1, i, i+1)$ between the lines $(i-1, i)$ and $(i, i+1)$. This corresponds to a point x that is integrated along the edge of the space-time Wilson Loop between x_i and x_{i+1}

be reformulated as follows,

$$(\bar{\varepsilon} \cdot \bar{Q})\, \mathrm{W}(\mathcal{Z}_1, \ldots, \mathcal{Z}_n) = a \sum_i \int_{I \times S^1} V \lrcorner \mathrm{D}^{3|4} \mathcal{Z}\, \mathrm{W}(\mathcal{Z}_1, \ldots, \mathcal{Z}_i, \mathcal{Z}, \mathcal{Z}_{i+1}, \ldots, \mathcal{Z}_n)\,, \tag{6.52}$$

where the right-hand side involves integrating out a momentum twistor from an $(n+1)$ point superloop. The holomorphic vector field

$$V = \bar{\epsilon}_a^{\dot{\alpha}} \chi^a \frac{\partial}{\partial \mu^{\dot{\alpha}}} \tag{6.53}$$

generates self-dual $\bar{Q}$ supersymmetry transformations on twistor space, while $\mathrm{D}^{3|4}\mathcal{Z}$ is the holomorphic measure on $\mathbb{CP}^{3|4}$. The contraction can be written in terms of the components of the additional momentum twistor as

$$V \lrcorner \mathrm{D}^{3|4} \mathcal{Z} \equiv (\bar{\epsilon}_a, \mathrm{Z}, \mathrm{dZ}, \mathrm{dZ})\, \mathrm{d}^3 \chi^a. \tag{6.54}$$

The fermionic integral is performed by treating the new fermionic component χ^a as independent of the others χ_i^a. Bosonically, the additional twistor is constrained to lie in the plane $(i-1, i, i+1)$ and so may be parametrized as

$$Z = Z_i + p(Z_{i-1} + q Z_{i+1}) \tag{6.55}$$

whereupon the bosonic part of the measure becomes $(\bar{\epsilon}_a, i-1, i, i+1)\, \mathrm{d}q\, p \mathrm{d}p$. With this parametrization, the contour $S^1 \times I$ extracts the residue of the integrand at $p = 0$, and integrates q from 0 to ∞ (Fig. 6.3).

Geometrically, the $(n+1)$-point superloop is integrated over a contour that fixes the additional twistor $\mathcal{Z} \to \mathcal{Z}_i$, and then integrates the remaining line X in the plane $(i-1, i, i+1)$ between the lines X_i and X_{i+1}. In space-time, this corresponds to taking a particular collinear limit where the additional cusp is sent to the line $[x_i, x_{i+1}]$ and then integrated along it.

The importance of Eq. (6.52) is that it provides a representation of the curvature insertion at the level of the external data. The formula is recursive in the number of loops; the derivative on the left lowers the transcendentality by one, while on the right the integral over the contour I with boundary increases the transcendentality by one. Both the S^1 contour and the Grassmann integration preserve transcendentality. On the other hand, the operator on the left increases the Grassmann degree by one, while the Grassmann integration on the right decreases it by three. Consequently, contributions to the $\bar{Q}$ anomaly of an N^{k}MHV superamplitude arise from an N^{k+1}MHV superamplitude.

6.4.2 Abelian Theory

Let us gain some familiarity with the descent equation by reproducing the result of Sect. (6.3.5) for the superloop in the abelian theory and with framing regularization. Expanding to order g^2 in the coupling and order χ in the fermions, the descent equation gives

$$(\bar{\varepsilon}\cdot\bar{Q})\, \mathrm{W}^{(1)}_{n,0}(\mathcal{Z}_1, \ldots, \mathcal{Z}_n) = a \sum_i \int V\lrcorner \mathrm{D}^{3|4}\mathcal{Z}\, \mathrm{W}^{(0)}_{n+1,1}(\ldots, i, \mathcal{Z}, i{+}1, \ldots)\,, \quad (6.56)$$

For the tree-level NMHV superamplitude, the framed Wilson loop is given by

$$\mathrm{W}^{(0)}_{n+1,1} = \frac{1}{2}\sum_{k,j} [*, i, i+1, j, j+1] \quad (6.57)$$

corresponding to a single propagator stretched between the two copies of the framed superloop in twistor space [28, 32]. Here, the indices i and j each run around one of the two copies of the $(n+1)$-point superloop, respectively. We will assume for simplicity that the reference supertwistor has no fermionic component, $\mathcal{Z}_* = (Z_*, 0)$; any contributions from a non-vanishing fermionic component χ_* may be verified to cancel around the sum.

The only dual superconformal invariants in this sum that have a pole as $\mathcal{Z} \to \mathcal{Z}_i$ are $[*, i, \mathcal{Z}, j, j+1]$ for some j. We thus find

$$\sum_i \int V\lrcorner \mathrm{D}^{3|4}\mathcal{Z}\, \mathrm{W}^{(0)}_{n+1,1}(\ldots, i, \mathcal{Z}, i+1, \ldots) = \sum_{i,j} \int V\lrcorner \mathrm{D}^{3|4}\mathcal{Z}\, [*, i, \mathcal{Z}, j, j+1]$$

$$= \sum_{i,j} \int (\bar{\epsilon}_a, Z, \mathrm{d}Z, \mathrm{d}Z)(i, j, j+1, *)^2 \, \frac{\chi^a_i (Z, j, j+1, *) + \text{cyclic}}{(i, Z, j, j+1)\,(Z, j, j+1, *)\,(j+1, *, i, Z)\,(*, i, Z, j)}$$

$$= \sum_{i,j} (\bar{\epsilon}_a, i-1, i, i+1)(i, j, j+1, *) \int dq \, \frac{\chi_i^a(Z(q), j, j+1, *) + \text{cyclic}}{(i, Z(q), j, j+1)(j+1, *, i, Z(q))(*, i, Z(q), j)} \tag{6.58}$$

where in going to the second line we have performed the Grassmann integral, and in going to the third we used the explicit parametrization (6.55) and performed the p contour integral, expressing the answer in terms of $Z(q) = Z_{i-1} + qZ_{i+1}$.

The numerator of (6.58) contains terms proportional to $\chi_i, \chi_j, \chi_{j+1}$ and, via $\chi(q)$ also terms proportional to the fermions χ_{i-1} and χ_{i+1}. After relabeling $j \to j-1$ in the term proportional to χ_{j+1} we find a contribution

$$\sum_{i,j} \int dq \, \frac{(\bar{\epsilon}_a, i-1, i, i+1)\,(i, j-1, j, j+1)}{(i, Z(s), j, j+1)\,(i, Z(s), j-1, j)} \chi_j^a = \sum_{i,j} \frac{(\bar{\epsilon}_a, i-1, i, i+1)}{(j, i-1, i, i+1)} \chi_j^a \log \frac{X_i \cdot Y_{j+1} \; X_{i+1} \cdot Y_j}{X_i \cdot Y_j \; X_{i+1} \cdot Y_{j+1}}. \tag{6.59}$$

For the term $\propto \chi_i$ we first write the numerator $(\bar{\epsilon}_a, i-1, i, i+1)(Z(q), j, j+1, *)$ using a four-term identity,

$$(\bar{\epsilon}_a, i-1, j, i+1)(Z(q), i, j+1, *) + (\bar{\epsilon}_a, i-1, j+1, i+1)(Z(q), j, i, *) \\ +(\bar{\epsilon}_a, i-1, *, i+1)(Z(q), j, j+1, i)$$

and note that the final term cancels telescopically in the sum over j. Now relabeling $j \to j-1$ in the second term here we find a contribution

$$\sum_{i,j} \int dq \, \frac{(j, i-1, \bar{\epsilon}_a, i+1)\,(i, j-1, j, j+1)}{(i, Z(s), j, j+1)\,(i, Z(s), j-1, j)} \chi_j^a = \sum_{i,j} \frac{(j, i-1, \bar{\epsilon}_a, i+1)}{(j, i-1, i, i+1)} \chi_i^a \log \frac{X_i \cdot Y_{j+1} \; X_{i+1} \cdot Y_j}{X_i \cdot Y_j \; X_{i+1} \cdot Y_{j+1}}. \tag{6.60}$$

After noting that

$$(\bar{\epsilon}_a, i-1, i, i+1) = (\bar{\epsilon}_a, Z(q), i, i+1) = \frac{1}{q}(\bar{\epsilon}_a, i-1, i, Z(q)), \tag{6.61}$$

the terms proportional to the fermions χ_{i-1} and χ_{i+1} may be treated similarly. Combining all the terms gives

$$\bar{Q}\, \mathrm{W}_{\mathrm{MHV}}^{1-\mathrm{loop}}[C_n] = \sum_{i,j} \log \frac{X_i \cdot Y_{j+1} \; X_{i+1} \cdot Y_j}{X_i \cdot Y_j \; X_{i+1} \cdot Y_{j+1}} \; \bar{Q} \log(j, i-1, i, i+1) \; + \; (i \leftrightarrow j) \tag{6.62}$$

and it follows that the symbol of the framed one-loop MHV amplitude is given by

$$\mathrm{dW}_{\mathrm{MHV}}^{1-\mathrm{loop}}[C_n] = \sum_{i,j} \log \frac{X_i \cdot X_{j+1}\; X_{i+1} \cdot X_j}{X_i \cdot X_j\; X_{i+1} \cdot X_{j+1}}\, \mathrm{d}\log(i{-}1, i, i{+}1, j) + (i \leftrightarrow j) \quad (6.63)$$

in agreement with our previous result (6.51).

6.4.3 Derivation

Our task is now to show that the descent equation reproduces the space-time Ward identity derived in section. This demonstration depends critically on the relationship between the holomorphic twistor space Wilson loop and space-time superloop, which is provided by the holomorphic frames $H(x, \lambda)$. The holomorphic parallel propagator is constructed from the combination

$$\mathrm{U}_{X_j}(\lambda_j, \lambda_{j-1}) = H(x_j, \lambda_j) H^{-1}(x_j, \lambda_{j-1}) \quad (6.64)$$

whereas the space-time parallel propagator is

$$\mathrm{P}\,\exp\,\mathrm{i} \int_{x_j}^{x_{j+1}} \mathbb{A} = H^{-1}(x_{j+1}, \lambda_j) H(x_j, \lambda_j)\,. \quad (6.65)$$

Thus, pairing the holomorphic frames differently around the curve as

$$\cdots H(x_{i+1}; \lambda_{i+1})\, \underbrace{\overbrace{H^{-1}(x_{i+1}, \lambda_i)\, H(x_i, \lambda_i)}^{}}_{\mathrm{P}\exp\mathrm{i}\int_{x_i}^{x_{i+1}}\mathbb{A}}\, \underbrace{H^{-1}(x_i, \lambda_{i-1})\, H(x_{i-1}, \lambda_{i-1})}_{\mathrm{P}\exp\mathrm{i}\int_{x_{i-1}}^{x_i}\mathbb{A}} \cdots,$$

where the overbraces group $\mathrm{U}_{\mathrm{X}_{i+1}}(\mathcal{Z}_{i+1}, \mathcal{Z}_i) = H(x_{i+1}; \lambda_{i+1})\, H^{-1}(x_{i+1}, \lambda_i)$ and $\mathrm{U}_{\mathrm{X}_i}(\mathcal{Z}_i, \mathcal{Z}_{i-1}) = H(x_i, \lambda_i)\, H^{-1}(x_i, \lambda_{i-1})$

(6.66)

both the twistor and space-time superloops may be exhibited simultaneously. This observation forms the cornerstone of the argument that follows.

Now let us apply the above technique to re-express the right hand side of the $\bar{Q}$ Ward identity in a more convenient form. We find that the argument of the insertion formula becomes

$$\begin{aligned}
&\mathrm{Tr}\,\mathrm{P}\left[\cdots \exp\left(\mathrm{i}\int_x^{x_{i+1}} \mathbb{A}\right) [\bar{\varepsilon}_a|\mathrm{d}x|\mathcal{F}^a\rangle \exp\left(\mathrm{i}\int_{x_i}^{x} \mathbb{A}\right) \cdots\right] \\
&= \mathrm{Tr}\,\mathrm{P}\Big[\cdots H^{-1}(x_{i+1}, \lambda_i)\, H(x, \lambda_i)\, [\bar{\varepsilon}_a|\mathrm{d}x|\mathcal{F}^a\rangle\, H^{-1}(x, \lambda_i)\, H(x_i, \lambda_i) \cdots\Big] \qquad (6.67) \\
&= \mathrm{Tr}\Big[\cdots \mathrm{U}_{\mathrm{X}_{i+1}}(\mathcal{Z}_{i+1}, \mathcal{Z}_i)\, H(x, \lambda_i)\, [\bar{\varepsilon}_a|\mathrm{d}x|\mathcal{F}^a\rangle\, H^{-1}(x, \lambda_i)\, \mathrm{U}_{\mathrm{X}_i}(\mathcal{Z}_i, \mathcal{Z}_{i-1}) \cdots\Big]
\end{aligned}$$

so that all the holomorphic frames except those immediately adjacent to the insertion may be paired into holomorphic parallel propagators.

Let us now derive this insertion formula from the descent equation (6.52). The $(n+1)$-point superloop can be expressed terms of the expectation value of the product of holomorphic frames

$$\cdots \mathrm{U}(\mathcal{Z}_{i+1}, \mathcal{Z})\mathrm{U}(\mathcal{Z}, \mathcal{Z}_i)\mathrm{U}(\mathcal{Z}_i, \mathcal{Z}_{i-1}) \cdots . \tag{6.68}$$

In the contour integral over the additional supertwistor $\mathcal{Z}$, the only terms that contribute are those leading to double poles $1/p^2$ once the fermionic components χ^a have been integrated out. The only such contributions come from the fermionic integral $d^4\chi$ acting on the holomorphic parallel propagator $\mathrm{U}(\mathcal{Z}, i)$. Thus, any field insertions get trapped between momentum twistors $\mathcal{Z}_i$ and $\mathcal{Z}$ when we take the residue at $s = 0$ and consequently send $\mathcal{Z} \to \mathcal{Z}_i$.

Let us now consider this argument in more detail. Pulling back to the spin bundle over the line X, the bosonic coordinates of $\mathcal{Z}$ are $(\lambda_\alpha, \mathrm{i}x^{\alpha\dot{\alpha}}\lambda_\alpha)$ and the bosonic measure in Eq. (6.54) becomes

$$(\bar{\epsilon}_a, \mathrm{Z}, \mathrm{dZ}, \mathrm{dZ}) = [\bar{\epsilon}_a \mathrm{d}\mu] \, \langle \lambda \mathrm{d}\lambda \rangle = \mathrm{i}[\bar{\epsilon}_a|\mathrm{d}x|\lambda\rangle \, \langle \lambda \mathrm{d}\lambda \rangle. \tag{6.69}$$

To compute the fermionic integration on the holomorphic parallel propagator $\mathrm{U}(\mathcal{Z}, i)$, we must replace the fermionic integrals $\mathrm{d}\chi \equiv \partial/\partial\chi$ by an operation on the θs, since $\mathrm{U}_\mathrm{X}(\mathcal{Z}, i)$ depends on these fermions only through the combination $\chi^a = \theta^{\alpha a}\lambda_\alpha$. If we pull back a holomorphic function $f(\mathcal{Z})$ on super twistor space to the spin bundle, then

$$\left.\left(\frac{\partial f}{\partial \chi^a}\right)\right|_{\chi=\theta\lambda} = \frac{1}{\langle \rho\lambda \rangle} \rho^\alpha \frac{\partial f|_{\chi=\theta\lambda}}{\partial \theta^{\alpha a}} \tag{6.70}$$

where ρ^α is a reference spinor that defines a lift of the fermionic vector field $\partial/\partial\chi^a$ to the spin bundle. Here there is dictated by the problem in hand: since by definition

$$\theta^{\alpha a} = \frac{(\chi_i^a \lambda^\alpha - \chi^a \lambda_i^\alpha)}{\langle \lambda, i \rangle}, \tag{6.71}$$

we must choose the lift $\rho^\alpha = \lambda_i^\alpha$ to ensure that we do not pick up contributions from the independent fermion $\chi_i^a = \theta^a|i\rangle$. The fermionic part of the integration can then be written

$$\int \varepsilon^{abcd} \frac{\langle i\, \partial_b \rangle \langle i\, \partial_c \rangle \langle i\, \partial_d \rangle}{\langle \lambda\, i \rangle^3} . \tag{6.72}$$

where we used the notation $\langle i \partial_a \rangle \equiv \lambda_i^\alpha \partial/\partial\theta^{\alpha a}$. This completes our discussion of the measure.

Let us now compute the fermionic derivative $\partial_{\alpha a} \equiv \partial/\partial\theta^{\alpha a}$ of the holomorphic parallel propagator $\mathrm{U}_\mathrm{X}(\mathcal{Z}, i)$. From its defining equation, the holomorphic parallel propagator obeys

$$
\begin{aligned}
0 &= \partial_{\alpha a}\Big[\, (\bar{\partial} - \mathrm{i}\mathcal{A})|_{\mathrm{X}}\, \mathrm{U}_{\mathrm{X}}(\mathcal{Z}, \mathcal{Z}_i) \,\Big] \\
&= (\bar{\partial} - \mathrm{i}\mathcal{A})\, \partial_{\alpha a}\mathrm{U}_{\mathrm{X}} - \mathrm{i}\lambda_\alpha \frac{\partial \mathcal{A}}{\partial \chi^a}\, \mathrm{U}_{\mathrm{X}}\,,
\end{aligned}
\tag{6.73}
$$

where in the second line we understand that all objects are pulled back to the spin-bundle at the point X. This equation is solved by inverting the $(\bar{\partial} - \mathrm{i}\mathcal{A})$-operator, and can be written in terms of the integral equation

$$
\partial_{\alpha a}\mathrm{U}_{\mathrm{X}}(\mathcal{Z}, \mathcal{Z}_i) = \mathrm{i}\int_{\mathrm{X}} \frac{\langle\lambda' \mathrm{d}\lambda'\rangle\langle\lambda\, i\rangle}{\langle\lambda\lambda'\rangle\langle\lambda' i\rangle} \mathrm{U}_{\mathrm{X}}(\mathcal{Z}, \mathcal{Z}')\, \lambda'_\alpha \frac{\partial \mathcal{A}}{\partial \chi'^a}\, \mathrm{U}_{\mathrm{X}}(\mathcal{Z}', \mathcal{Z}_i)
\tag{6.74}
$$

over another point $\mathcal{Z}' \in \mathrm{X}$.

Combining the expressions for the measure (6.69), (6.72) and the integrand (6.74) we see that the right hand side of the descent equation may be written as

$$
\begin{aligned}
&\int V \lrcorner \mathrm{D}^{3|4}\mathcal{Z}\; \mathrm{W}(\cdots, i, \mathcal{Z}, i+1, \cdots) \\
&= -\frac{\mathrm{i}}{3!N}\int [\bar{\epsilon}_a|\mathrm{d}x|\lambda\rangle\langle\lambda \mathrm{d}\lambda\rangle\, \mathrm{Tr}\Big[\cdots \varepsilon^{abcd}\frac{\langle i\partial_b\rangle\langle i\partial_c\rangle}{\langle i\lambda\rangle^2}\int_{\mathrm{X}}\frac{\langle\lambda' \mathrm{d}\lambda'\rangle}{\langle\lambda\lambda'\rangle}\mathrm{U}_{\mathrm{X}}(\mathcal{Z}, \mathcal{Z}')\, \frac{\partial \mathcal{A}}{\partial \chi'^d}\, \mathrm{U}_{\mathrm{X}}(\mathcal{Z}', \mathcal{Z}_i)\cdots\Big] \\
&= -\frac{\mathrm{i}}{3!N}\int_{x_i}^{x_{i+1}} \mathrm{d}t \oint \langle\lambda \mathrm{d}\lambda\rangle\, [\bar{\epsilon}_a i]\, \mathrm{Tr}\Big[\cdots \varepsilon^{abcd}\frac{\langle i\partial_b\rangle\langle i\partial_c\rangle}{\langle i\lambda\rangle}\int_{\mathrm{X}}\frac{\langle\lambda' \mathrm{d}\lambda'\rangle}{\langle\lambda\lambda'\rangle}\mathrm{U}_{\mathrm{X}}(\mathcal{Z}, \mathcal{Z}')\, \frac{\partial \mathcal{A}}{\partial \chi'^d}\, \mathrm{U}_{\mathrm{X}}(\mathcal{Z}', \mathcal{Z}_i)\cdots\Big],
\end{aligned}
\tag{6.75}
$$

where in the last line we have used the parametrization $\mathrm{d}x = |i\rangle[i|\,\mathrm{d}t$. It is now straightforward to perform the contour integral using the simple pole in $\langle i\lambda\rangle$, which leads to the expression

$$
-\frac{\mathrm{i}}{N}\int_{x_i}^{x_{i+1}} \mathrm{d}t\, [\bar{\epsilon}_a \tilde{\imath}]\, \mathrm{Tr}\Big[\cdots \frac{\varepsilon^{abcd}}{3!}\langle i\partial_b\rangle\langle i\partial_c\rangle \int_{\mathrm{X}}\frac{\langle\lambda' \mathrm{d}\lambda'\rangle}{\langle i\lambda'\rangle}\mathrm{U}_{\mathrm{X}}(\mathcal{Z}_i, \mathcal{Z}')\, \frac{\partial \mathcal{A}}{\partial \chi'^d}\, \mathrm{U}_{\mathrm{X}}(\mathcal{Z}', \mathcal{Z}_i)\cdots\Big].
\tag{6.76}
$$

We would now like to relate this expression to the insertion of a local operator in space-time, using the non-linear Penrose transform.

Let us now recall that the fermionic space-time superconnection is obtained from the holomorphic frames by

$$
\mathrm{i}\lambda^\alpha \mathcal{A}_{\alpha a}(x, \theta) = H_{\mathrm{X}}^{-1}\lambda^\alpha \partial_{\alpha a}\, H_{\mathrm{X}}\,.
\tag{6.77}
$$

This equation allows the fermionic superconnection to be reconstructed up to super-gauge transformations. The super gauge transformations act on the holomorphic frames in the following way, $H_{\mathrm{X}} \to H_{\mathrm{X}}\, g_{\mathrm{X}}$. Here we will perform a super gauge transformation by $g_{\mathrm{X}} = H_{\mathrm{X}}^{-1}(\lambda_i)$ so that in this gauge the fermionic superconnection is

$$
\mathrm{i}\lambda^\alpha \mathcal{A}_{\alpha a}(x, \theta, \lambda_i) = \mathrm{U}_{\mathrm{X}}(\lambda_i, \lambda)\lambda^\alpha \partial_{\alpha a}\, \mathrm{U}(\lambda, \lambda_i)\,.
\tag{6.78}
$$

where the holomorphic parallel propagators normalized to be the identity when $\lambda = \lambda_i$. Evaluating the expression when $\lambda = \lambda_i$ we immediately find that $\lambda_i^\alpha \Gamma_{\alpha a}(x, \theta) = 0$, and hence in this gauge

$$\mathcal{A}_{\alpha a}(x, \theta) = \lambda_{\alpha i} \mathcal{A}_a(x, \theta; \lambda_i) \tag{6.79}$$

where $\mathcal{A}_a$ is a fermionic Lorentz scalar that depends smoothly on the choice of spinor λ_i. Now using the formula (6.74) for the fermionic derivative of a holomorphic frame we find

$$\mathcal{A}_a(x, \theta; \lambda_i) = \int_{\mathrm{X}} \frac{\langle \lambda' \mathrm{d}\lambda' \rangle}{\langle i \lambda' \rangle} \mathrm{U}_{\mathrm{X}}(\mathcal{Z}_i, \mathcal{Z}') \frac{\partial \mathcal{A}}{\partial \chi'^a} \mathrm{U}_{\mathrm{X}}(\mathcal{Z}', \mathcal{Z}_i) \,, \tag{6.80}$$

which is precisely the insertion as appears in (6.76).

The existence of a gauge in which $|\mathcal{A}_a\rangle = |i\rangle \mathcal{A}_a$ has a remarkable consequence. From the integrability conditions (6.23), the only non-vanishing part of the fermionic-fermionic supercurvature is

$$\begin{aligned} \mathcal{F}_{ab} &= \frac{\mathrm{i}}{2} \epsilon^{\alpha\beta} \left\{ \nabla_{\alpha a}, \nabla_{\beta b} \right\} \\ &= \partial^\alpha_{\ [a} \mathcal{A}_{b]\alpha} - \mathrm{i} \{ \mathcal{A}^\alpha_{\ a}, \mathcal{A}_{\alpha b} \} \end{aligned} \tag{6.81}$$

but if the fermionic superconnection $\Gamma_{\alpha a}$ is proportional to a fixed spinor then the final anti-commutator vanishes, even in the non-Abelian theory. Therefore, in this gauge we have the following results for the fermionic-fermionic supercurvature

$$\mathcal{F}_{cd} = \partial^\alpha_{\ [c} \mathcal{A}_{d]\alpha} = \langle i \partial_{[c} \rangle \mathcal{A}_{d]} \tag{6.82}$$

and hence

$$\begin{aligned} \lambda_i^\alpha \mathcal{F}^a_{\ \alpha} &= \frac{1}{3!} \varepsilon^{abcd} \lambda_i^\alpha \nabla_{\alpha b} \mathcal{W}_{cd} \\ &= \frac{1}{3!} \varepsilon^{abcd} \langle i \partial_b \rangle \langle i \partial_c \rangle \mathcal{A}_d . \end{aligned} \tag{6.83}$$

Hence we have derived

$$\begin{aligned} &\int V \lrcorner \mathrm{D}^{3|4} \mathcal{Z}\, \mathrm{W}(\cdots, i, \mathcal{Z}, i{+}1, \cdots) \\ &= -\frac{\mathrm{i}}{N} \left\langle \int_{x_i}^{x_{i+1}} \mathrm{Tr} \left[\cdots \mathrm{U}_{\mathrm{X}_{i+1}}(\mathcal{Z}_{i+1}, \mathcal{Z}_i) \, [\bar{\epsilon}_a | \mathrm{d}x | \mathcal{F}^a \rangle \, \mathrm{U}_{\mathrm{X}_i}(\mathcal{Z}_i, \mathcal{Z}_{i-1}) \cdots \right] \right\rangle \end{aligned} \tag{6.84}$$

in the gauge where $|\mathcal{A}_a\rangle = |i\rangle \mathcal{A}_a(x, \theta; \lambda_i)$.

At present, the supercurvature insertion $|\mathcal{F}^a\rangle$ is expressed in a gauge that is very obscure from the space-time point of view. Thus, we should now transform back to the 'space-time' gauge where the twistor connection is harmonic, $(\bar{\partial}^{\dagger}\mathcal{A})_{\mathrm{X}} = 0$, with respect to an arbitrary Hermitian metric on the line X [33, 34]. We can assume that the basic holomorphic frame $\mathrm{H}(x, \theta, \lambda)$ has been in this gauge from the beginning. Then, performing the super gauge transformation

$$\mathrm{U}_{\mathrm{X}}(\lambda, \lambda_i) \to \mathrm{U}_{\mathrm{X}}(\lambda, \lambda_i)\,\mathrm{H}^{-1}(x, \theta; \lambda_i) = \mathrm{H}(x, \theta; \lambda) \tag{6.85}$$

then the supercurvature insertion transforms as

$$\mathcal{F}^a_\alpha \to \mathrm{H}(x, \theta, \lambda_i)\,\mathcal{F}^a_\alpha\,\mathrm{H}^{-1}(x, \theta; \lambda_i). \tag{6.86}$$

Hence we now have

$$\frac{\mathrm{i}g^2}{N}\left\langle \oint \mathrm{Tr}\big([\bar{\epsilon}_a|\mathrm{d}x|\mathcal{F}^a\rangle\;\mathrm{Hol}[C]\big)\right\rangle \tag{6.87}$$

as required.

6.4.4 Ratio Function

Let us now address one subtlety in the above derivation. Although the curvature insertion formula (6.38) is proportional to the coupling g^2, we believe that the right hand side of the descent relation (6.52) between superloops or superamplitudes should be proportional to the cusp anomalous dimension $\Gamma(a)$ in the non-abelian theory. This is because of renormalization of the curvature insertion in the collinear limit [26].

An alternative way to see that the right-hand side is proportional to $\Gamma(a)$ is to derive a consistent anomaly equation for the finite remainder function $\mathcal{R}_n$. Here we use the following representation of the remainder function [35]

$$\mathcal{R}_n \equiv \mathrm{W}[C_n] \,/\, \mathrm{W}_{\mathrm{ab}}[C_n]^{\Gamma(a)/g^2} \tag{6.88}$$

where $\mathrm{W}_{\mathrm{ab}}[C_n]$ is the regular bosonic loop in the abelian theory. This requires an adjustment of the regulator in the abelian theory in order to remove regulator dependent collinear divergences. Since the external $\bar{Q}$ operator obeys the Leibnitz rule we have

$$\begin{aligned}\bar{Q}\,\mathcal{R}_n &= \Gamma(a)\sum_i \int V \lrcorner \mathrm{D}^{3|4}\mathcal{Z}\,\left(\frac{\mathrm{W}[C_{n+1}(\mathcal{Z})]}{\mathrm{W}_{\mathrm{ab}}[C_n]^{\Gamma(a)/g^2}} - \mathcal{R}_n \mathrm{W}^{(0)}_{n+1,1}[C_{n+1}(\mathcal{Z})]\right)\\ &= \Gamma(a)\sum_i \int V \lrcorner \mathrm{D}^{3|4}\mathcal{Z}\,\Big(\mathcal{R}_{n+1}(\mathcal{Z}) - \mathcal{R}_n \mathcal{A}^{(0)}_{n+1,1}(\mathcal{Z})\Big)\,,\end{aligned} \tag{6.89}$$

In going to the second line we have used the fact that $\mathrm{W}_{\mathrm{ab}}[C_n]$ is independent of the fermionic momentum twistors, so that in the first term on the right, the pole in p is provided by the Grassmann integral acting solely on the numerator. Also, since the

BDS ansatz provides the correct behaviour under collinear limits, we have replaced the n-particle BDS ansatz by the $\mathcal{Z}$-dependent $(n+1)$-particle BDS ansatz under this contour integral.

6.5 Discussion and Outlook

The authors of [26] have provided significant evidence that the $\bar{Q}$ anomaly equation for the remainder function is indeed correct. In particular, the cancellation of potential divergences in the collinear integral between the terms on the right hand side requires that the overall coefficient be the cusp anomalous dimension. Nevertheless, our arguments certainly do not constitute a rigorous derivation, which would require a proper understanding of how to regulate the Wilson superloop in a way consistent with supersymmetry.

The dual superconformal $\bar{Q}$ anomaly is not enough to compute to compute the remainder function in general, even together with collinear limits. The reason is that the kernel becomes too complicated beyond N^{2}MHV. In general, one also needs the anomaly equation for the level-one Yangian generator $Q^{(1)}$. This has been derived by a parity transformation in [26] but the bi-local operator appears too complicated to gain any useful information, at least with our present level of understanding. Nevertheless, the $Q^{(1)}$ anomaly equation deserves further investigation.

Finally, the anomaly equation (6.89) for the remainder function should be correct also at strong coupling. In particular, the cusp anomalous dimension is known to all values of the coupling thanks to integrability [36]. Unfortunately, our understanding of scattering amplitudes at strong coupling is extremely limited beyond the MHV sector, where much is known [37–39]. An important goal for the future is to extend such integrability techniques beyond the MHV sector. We hope that progress will soon be made in this direction and that the anomaly equation can be useful in the strong coupling regime.

Appendix: Component Expansions

The superconnection $\mathbb{A}$ on chiral superspace is determined by the superspace constraints and supergauge condition $\theta^{\alpha a}\mathcal{A}_{\alpha a} = 0$. Up to order fourth order in the fermions, the constraints are solved by the following expansion

$$\begin{aligned}\mathcal{A} &= A + \mathrm{i}|\theta^a\rangle[\bar{\psi}_a| + \frac{\mathrm{i}}{2}|\theta^a\rangle\langle\theta^b|D\phi_{ab} - \frac{1}{3!}\varepsilon_{abcd}|\theta^a\rangle\langle\theta^b|\,D\langle\theta^c\psi^d\rangle \\ &\quad + \frac{\mathrm{i}}{4!}\varepsilon_{abcd}|\theta^a\rangle\langle\theta^b|\,D\langle\theta^c|G|\theta^d\rangle + \cdots \qquad (6.90)\\ |\mathcal{A}_a\rangle &= \frac{\mathrm{i}}{2}\phi_{ab}|\theta^b\rangle - \frac{1}{3}\varepsilon_{abcd}|\theta^b\rangle\langle\theta^c\psi^d\rangle + \frac{\mathrm{i}}{8}\varepsilon_{abcd}|\theta^b\rangle\langle\theta^c|G|\theta^d\rangle + \cdots .\end{aligned}$$

The corresponding supercurvatures have component expansions up to second order in the fermions as follows

$$
\begin{aligned}
\mathcal{F}^{+}_{\dot\alpha\dot\beta}(x,\theta) &= F^{+}_{\dot\alpha\dot\beta} + \mathrm{i}\langle\theta^{a}|D_{(\dot\alpha}\bar\psi_{\dot\beta)a} + \frac{\mathrm{i}}{2}\langle\theta^{a}|D_{(\dot\alpha}\langle\theta^{b}|D_{\dot\beta)}\phi_{ab} - \frac{\mathrm{i}}{2}\langle\theta^{a}\theta^{b}\rangle\left\{\bar\psi_{a(\dot\alpha},\bar\psi_{\dot\beta)b}\right\} + \cdots \\
\mathcal{F}^{-}_{\alpha\beta}(x,\theta) &= F^{-}_{\alpha\beta} + \mathrm{i}\theta^{a}_{(\alpha}D_{\beta)\dot\beta}\bar\psi^{\dot\beta}_{a} + \mathrm{i}\theta^{\;a}_{(\alpha}\theta^{\;b}_{\beta)}\left(\Box\phi_{ab} - \left\{\bar\psi^{\dot\alpha}_{a},\bar\psi_{\dot\alpha b}\right\}\right) + \frac{1}{4}\theta^{\gamma b}\theta^{a}_{(\alpha}F^{-}_{\beta)\gamma}\phi_{ab} + \cdots \\
\mathcal{F}_{\dot\alpha a}(x,\theta) &= \bar\psi_{\dot\alpha a} + \theta^{\alpha b}D_{\alpha\dot\alpha}\phi_{ab} + \frac{\mathrm{i}\varepsilon_{abcd}}{3!}\theta^{\alpha b}D_{\alpha\dot\alpha}\langle\theta^{c}\psi^{d}\rangle + \cdots \\
\mathcal{W}_{ab}(x,\theta) &= \phi_{ab} + \mathrm{i}\varepsilon_{abcd}\langle\theta^{c}\psi^{d}\rangle + \frac{1}{2}\varepsilon_{abcd}\langle\theta^{c}|G|\theta^{d}\rangle + \frac{1}{4}\left[\phi_{ac},\phi_{bd}\right]\langle\theta^{c}\theta^{d}\rangle + \cdots .
\end{aligned}
\tag{6.91}
$$

See Refs. [32, 40] for further information.

References

1. J.M. Drummond, J.M. Henn, J. Plefka, Yangian symmetry of scattering amplitudes in N = 4 super Yang-Mills theory. JHEP **0905**, 046 (2009), [arXiv:0902.2987]
2. J. Drummond, L. Ferro, Yangians, Grassmannians and T-duality. JHEP **1007**, 027 (2010), [arXiv:1001.3348]
3. R. Akhoury, Mass divergences of wide angle scattering amplitudes. Phys. Rev. **D19**, 1250 (1979)
4. A.H. Mueller, On the asymptotic behavior of the sudakov form-factor. Phys. Rev. **D20**, 2037 (1979)
5. J.C. Collins, Algorithm to compute corrections to the Sudakov form-factor. Phys. Rev. **D22**, 1478 (1980)
6. A. Sen, Asymptotic behavior of the Sudakov form-factor in QCD. Phys. Rev. **D24**, 3281 (1981)
7. G.F. Sterman, Summation of large corrections to short distance Hadronic cross-sections. Nucl. Phys. **B281**, 310 (1987)
8. J. Botts, G.F. Sterman, Hard elastic scattering in QCD: leading behavior. Nucl. Phys. **B325**, 62 (1989)
9. S. Catani, L. Trentadue, Resummation of the QCD perturbative series for hard processes. Nucl. Phys. **B327**, 323 (1989)
10. G. Korchemsky, Sudakov form-factor in QCD. Phys. Lett. **B220**, 629 (1989)
11. G. Korchemsky, Double logarithmic asymptotics in QCD. Phys. Lett. **B217**, 330–334 (1989)
12. L. Magnea, G.F. Sterman, Analytic continuation of the Sudakov form-factor in QCD. Phys. Rev. **D42**, 4222–4227 (1990)
13. G. Korchemsky, G. Marchesini, Resummation of large infrared corrections using Wilson loops. Phys. Lett. **B313**, 433–440 (1993)
14. S. Catani, The singular behavior of QCD amplitudes at two loop order. Phys. Lett. **B427**, 161–171 (1998), [hep-ph/9802439]
15. G.F. Sterman, M.E. Tejeda-Yeomans, Multiloop amplitudes and resummation. Phys. Lett. **B552**, 48–56 (2003), [hep-ph/0210130]
16. G. Korchemsky, J. Drummond, E. Sokatchev, Conformal properties of four-gluon planar amplitudes and Wilson loops. Nucl. Phys. **B795**, 385–408 (2008), [arXiv:0707.0243]
17. J. Drummond, J. Henn, G. Korchemsky, E. Sokatchev, Conformal Ward identities for Wilson loops and a test of the duality with gluon amplitudes. Nucl. Phys. **B826**, 337–364 (2010), [arXiv:0712.1223]
18. I. Korchemskaya, G. Korchemsky, On lightlike Wilson loops. Phys. Lett. **B287**, 169–175 (1992)

19. A. Bassetto, I. Korchemskaya, G. Korchemsky, G. Nardelli, Gauge invariance and anomalous dimensions of a light cone Wilson loop in lightlike axial gauge. Nucl. Phys. **B408**, 62–90 (1993), [hep-ph/9303314]
20. S. Ivanov, G. Korchemsky, A. Radyushkin, Infrared asymptotics of perturbative QCD: contour gauges. Yad. Fiz. **44**, 230–240 (1986)
21. G. Korchemsky, G. Marchesini, Structure function for large X and renormalization of Wilson loop. Nucl. Phys. **B406**, 225–258 (1993), [hep-ph/9210281]
22. Z. Bern, L.J. Dixon, V.A. Smirnov, Iteration of planar amplitudes in maximally supersymmetric Yang-Mills theory at three loops and beyond. Phys. Rev. **D72**, 085001 (2005), [hep-th/0505205]
23. L.F. Alday, R. Roiban, Scattering amplitudes, Wilson loops and the String/Gauge theory correspondence. Phys. Rep. **468**, 153–211 (2008), [arXiv:0807.1889]
24. G. Korchemsky, E. Sokatchev, Symmetries and analytic properties of scattering amplitudes in N = 4 SYM theory. Nucl. Phys. **B832**, 1–51 (2010), [arXiv:0906.1737]
25. A. Sever, P. Vieira, *Symmetries of the N = 4 SYM S-matrix*, arXiv:0908.2437
26. S. Caron-Huot, S. He, Jumpstarting the all-loop S-matrix of planar N = 4 super Yang-Mills. JHEP **1207**, 174 (2012), [arXiv:1112.1060]
27. M. Bullimore, D. Skinner, *Descent Equations for Superamplitudes*, arXiv:1112.1056
28. M. Bullimore, D. Skinner, *Holomorphic Linking, Loop Equations and Scattering Amplitudes in Twistor Space*, arXiv:1101.1329
29. D. Gaiotto, J. Maldacena, A. Sever, P. Vieira, Pulling the straps of polygons. JHEP **1112**, 011 (2011), [arXiv:1102.0062]
30. S. Caron-Huot, Superconformal symmetry and two-loop amplitudes in planar N = 4 super Yang-Mills, arXiv:1105.5606 (Temporary entry)
31. A. Brandhuber, P. Heslop, G. Travaglini, MHV amplitudes in N = 4 super Yang-Mills and Wilson loops. Nucl. Phys. **B794**, 231–243 (2008), [arXiv:0707.1153]
32. L. Mason, D. Skinner, The complete planar S-matrix of N =4 SYM as a Wilson loop in Twistor space. JHEP **1012**, 018 (2010), [arXiv:1009.2225]
33. R. Boels, L. Mason, D. Skinner, Supersymmetric gauge theories in Twistor space. JHEP **0702**, 014 (2007), [hep-th/0604040]
34. N. Woodhouse, Real methods in Twistor theory. Class. Quant. Grav. **2**, 257–291 (1985)
35. L.F. Alday, D. Gaiotto, J. Maldacena, A. Sever, P. Vieira, An operator product expansion for polygonal null Wilson loops. JHEP **1104**, 088 (2011), [arXiv:1006.2788]
36. N. Beisert, B. Eden, M. Staudacher, Transcendentality and crossing. J. Stat. Mech. **0701**, P01021 (2007), [hep-th/0610251]
37. L.F. Alday, J. Maldacena, *Minimal Surfaces in AdS and the Eight-Gluon Scattering Amplitude at Strong Coupling*, arXiv:0903.4707
38. L.F. Alday, D. Gaiotto, J. Maldacena, Thermodynamic bubble Ansatz. JHEP **1109**, 032 (2011), [arXiv:0911.4708]
39. L.F. Alday, J. Maldacena, A. Sever, P. Vieira, Y-system for scattering amplitudes. J. Phys. A **A43**, 485401 (2010), [arXiv:1002.2459]
40. A. Belitsky, G. Korchemsky, E. Sokatchev, Are scattering amplitudes dual to super Wilson loops? Nucl. Phys. **B855**, 333–360 (2012), [arXiv:1103.3008]

Zeitfracht Medien GmbH
Ferdinand-Jühlke-Straße 7
99095 Erfurt, Deutschland
produktsicherheit@kolibri360.de